"What's this? *Another* book on Rubik's cube? Yes, but for ardent cubemeisters and all those who are interested in the mathematics of the cube, it was very much worth doing. It is a splendid book, clearly written, enlivened by delightful cartoons, and containing a wealth of information not to be found in any other book."

—**MARTIN GARDNER**

Ernő Rubik and Cube

CHRISTOPH BANDELOW

INSIDE RUBIK'S CUBE and BEYOND

with 21 cartoons by Alexander Mága

BIRKHÄUSER
BOSTON • BASEL • STUTTGART

CHRISTOPH BANDELOW
Haarholzer Strasse 13
D-4630 Bochum-Stiepel
FEDERAL REPUBLIC OF GERMANY

Translated by
Jeannette Zehnder and Lucy Moser

Library of Congress Cataloging in Publication Data
Bandelow, Christoph.
Inside Rubik's cube and beyond.
Translation of: Einführung in die Cubologie.
Bibliography: p.
Includes index.
1. Rubik's Cube. 2. Mathematical models.
3. Groups. Theory of. I. Title.
QA491.B2613 511'.8 82-1264
ISBN 3-7643-3078-3 AACR2

CIP-Kurztitelaufnahme der Deutschen Bibliothek
Bandelow, Christoph:
Inside Rubik's cube and beyond / Christoph
Bandelow. Transl. by Jeannette Zehnder and
Lucy Moser.—Boston ; Basel ; Stuttgart :
Birkhäuser, 1982.
Aus d. Ms. übers.
ISBN 3-7643-3078-3

ISBN 3-7643-3078-3
Printed in USA

PREFACE

On January 30, 1975 Ernő Rubik jr., professor of architecture and design in Budapest, was granted the Hungarian patent number 170062 for a "térbeli logikai játék"—a game of spatial logic. Between 1978 and March 1981 this object—Bűvös Kocka in Hungary, der Magische Würfel or Zauberwürfel in Germany, le Cube Hongrois in France and the Magic Cube or Rubik's Cube in Great Britain and the USA—has sold more than ten million copies. And they were not merely sold! A highly contagious "twist mania" has been spreading throughout families, offices and waiting rooms. Many classrooms sound as if an army of mice were hard at work behind the desks.

What is so fascinating about this cube, which competes with Hungarian salami and the famous Tokajer wine in the currency-winning export market? For one thing, it is an amazing technical tool. How does it work? Moreover, the contrast between its innocent, innocuous appearance and the hidden difficulty of its solution offers a serious challenge to all puzzle fans, but especially to those mathematicians who are professionally concerned with logical deduction.

But whatever the appearance—have no fear! **To enjoy playing with the cube and to study this book requires no previous mathematical knowledge.** Admittedly, an attempt is made in chapter two to develop the "mathematics of the cube" with uncompromising severity, as only *unadulterated* mathematics is *understandable* mathematics. However, everything is explained from basics. Besides, the reader who still feels overwhelmed can skim over this chapter without risking serious consequences. And there may even be a chance that—faced with such an unusually aesthetic object—one or two readers who so far regarded this field with trepidation or hostility will suddenly discover the "discreet charms of mathematics."

The educational benefit of our dealing with the cube is at least threefold: First, it gives an example of mathematical modeling; in other words, it describes a tiny piece of the real world by means of a theory. This is informative even for students of mathematics, who tend to neglect this exercise which is essential for the practical application of their subject. Second, many important ideas and theorems of group theory (and also some geometrical ones) can be examined by means of the cube in a natural and pleasant way which otherwise would be highly abstract and would leave no visual impression. The third benefit, finally, is the one stressed by Rubik himself: playing with the cube demands the capacity to imagine threedimensional space. For many people who have little practice in systematical thinking the complete mastery of a ready-made solution strategy with its hierarchy of case-distinctions means intellectual training.

Ready-made solution? Doesn't that spoil all the fun? This allegation cannot be categorically denied, of course. Developing a totally individual solution is great fun and gives a feeling of satisfaction. Only the person who has done this can judge the difficulty of the solitaire game. It is generally agreed that even shrewd puzzlers have to spend two weeks of constant fiddling, until they find a solution which in most cases is much too complicated. And not everybody is able or willing to invest so much time. Therefore, after giving a short description of the cube's technical properties and a system of notations, the book starts out with **a generally understandable and complete description of a simple solution strategy.** This, however, takes up only a few pages, since our main objective is totally different: **We want to understand the cube.** Chapter three enables every reader to develop his own individual strategy according to his own personal taste. Even more satisfying is perhaps that, after mastering certain basic maneuvers (f.i. some 3-cycles and double transpositions), you can get away from all those inflexible strategies and deal with any scrambled cube in your own way.

Among all the popular riddles, puzzles and "sliding games" Rubik's Cube is perhaps most likely to motivate its solver to develop his own little theory, part of a miniature science, which we have called, perhaps somewhat pompously, "cubology". Cubology today is still in its infancy. Many beautiful structural properties, generalizations, interesting analogies, effective new maneuvers or elegant strategies still await their discoverer. We hope to report on the progress of this search in a later edition of this book. The numeration of sections and figures was chosen such that improvements and insertions are easily possible.

For the sake of completeness we should mention something which might seem rather obvious: the study of this book is hardly possible without the cube in your hand; and make sure it is a good cube, not one of those cheap imitations, which can only be turned with difficulty. Also beware of cubes with round colored dots instead of inlaid colored tiles. These cubes limit the aesthetic attraction of otherwise very pretty positions.

When I began to work on this book shortly after acquiring my first cube, I thought I was doing a piece of original and independent work from beginning to end. Since then a whole series of articles on cubology has appeared (see bibliography), which I have eagerly absorbed. The booklet by David Singmaster has had the most enduring influence on me.

I am extremely grateful to Dr. Rolf Gall and Dipl.-Math. Norbert Krüger for proof-reading the manuscript and to cand. math. Helmut Corbeck for the final version of the figures. I would also like to express my thanks to the publishing firm, Friedrich Vieweg und Sohn, and especially to their reader, Dipl.-Math. Ulrike Schmickler-Hirzebruch, for their extremely pleasant and patient cooperation. My sons Tim and Nils, helped me enthusiastically in the systematic search for good maneuvers. In addition, Tim helped me with the fair copy of the manuscript, including the flow chart in chapter 6. Finally, I would like to thank my wife Dörte for her lovely cubological Christmas present, a truly magnanimous act in view of a totally "cube plagued" summer holiday. Her gift, a splendid cube stamp, is a great help for my quickly expanding maneuver card index.

DR. CHRISTOPH BANDELOW
Haarholzer Str. 13
D- 4630 Bochum 1

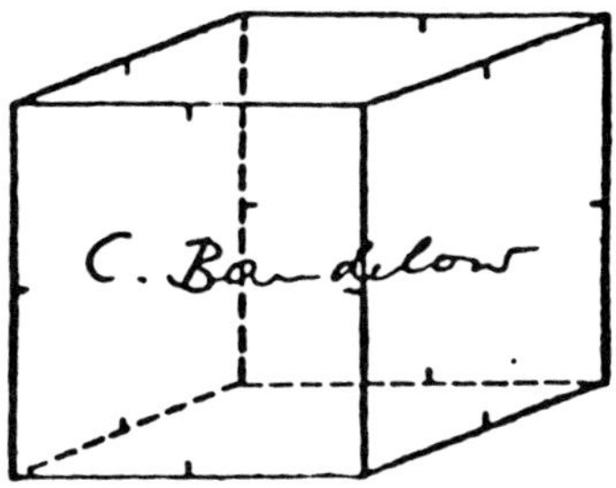

TABLE OF CONTENTS

1 SOLITAIRE

Avoiding all mathematics this Chapter attempts to look at the structure of the cube (Section 1.1), defines the basic concepts "maneuver" (1.2) and "operation" (1.5), and describes a simple strategy for solving the game of solitaire (1.3). In 1.4 we introduce a notation that is particularly appropriate for children and computers.

1.1 Cube, Cubies and Cubicles

Rubik's cube is made out of synthetic material. Each of its sides is approximately 5.6 cm long. It seems to consist of $3 \times 3 \times 3 = 27$ "cubies" of equal size. The $6 \times 9 = 54$ lateral faces of these cubies making up the surface of the cube are covered by colored tiles. At purchase, the cube is in its *start position:* Each of the 6 faces of the cube is one color. The 6 colors used on the faces of the cube are not the same for all cubes on the market: Those of Hungarian origin made out of black synthetics usually have yellow, white, blue, green, dark red and orange faces; while cubes made out of white synthetics often have one black face.

Each cubie is part of 3 different *layers* and each of these layers consists of 9 cubies. Thanks to an ingenious and amazingly simple mechanism each of the 9 layers can be turned freely in its plane around its center (figure 1). Four to five turns are usually enough to transform the one-colored faces of the start position into chaotic patchwork patterns. The object of the game is to recreate the start position out of this mix-up.

How is it possible to realize such an object technically? Why don't the cubies sitting at the corners simply fall off? They cannot after all be attached on any side because of the turning mechanism of the three

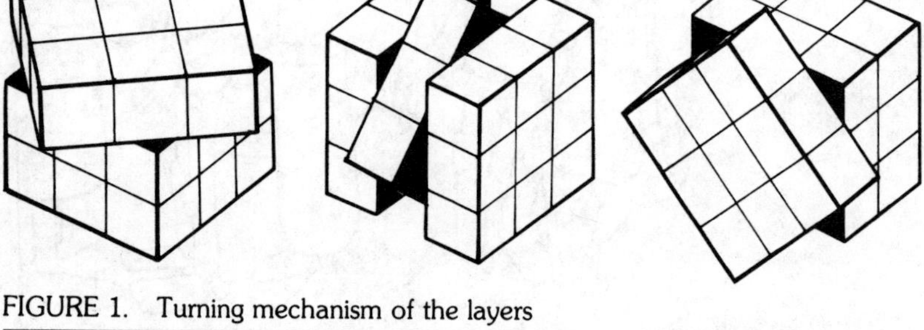

FIGURE 1. Turning mechanism of the layers

layers they belong to. It is easy to recognize three different types of cubies: 8 *corner cubies* with three colored tiles, 12 *edge cubies* with two colored tiles, and 6 *face cubies* with one colored tile. Each cubie remains true to type throughout all the turning of the layers. Photograph 1 shows how the technical aspects of the problem have been solved: The 27th cubie hidden in the interior of the cube is in reality a rigid system of three mutually perpendicular axes. The face cubies are actually half-cubies sitting flexibly and turnable at the tips of these axes. Since their position in relation to each other never changes (face cubies f.i. starting opposite each other will always be opposite each other), they determine the color of each face of the cube and the location of all the other cubies in the start position. Edge and corner cubies look more like cubes. The latter are equipped with a roundish sturdy "foot" by which they are firmly held in place by the three adjoining edge cubies. The edge cubies on their part each have in between the two gaps for the feet of the adjoining corner cubies a narrow oblong foot which is gripped by the two adjoining face cubies. Despite all the colorful goings-on, there is obviously firm control and the cube will not so easily fall apart. If we turn an outer layer we find that the four edge cubies each exchange one of their two corresponding face cubies, while the shared face cubie introduces his wards to their new guardians; the four corner cubies are each led by two supervising edge cubies towards their new, their third protector.

If you want to take a closer look at the inner workings of the cube, you can take your cube carefully apart without risking any damage. Turning f.i. the upper layer by 45° you can fairly easily remove an edge cubie while the face cubies of the underlying middle layer are just in the process of changing guards (figure 2).

Warning 1 (mathematical): In section 2.4 we shall prove that if we reassemble the cube randomly only every 12th position can be transformed

FIGURE 2.
Taking the Cube apart

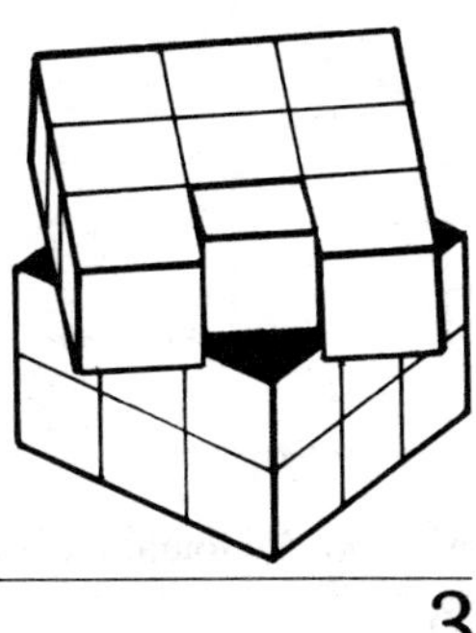

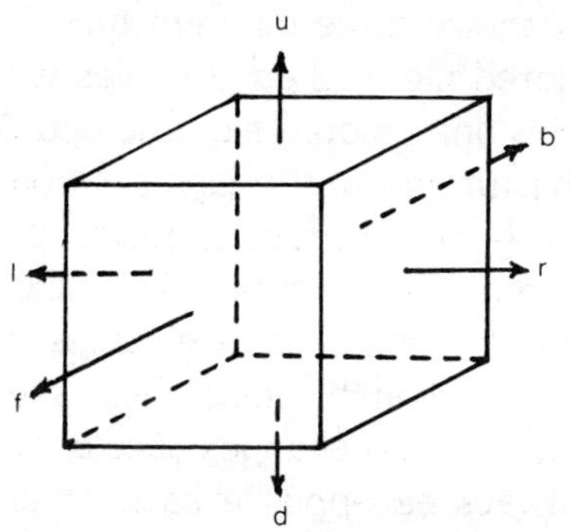

FIGURE 3.
Notation of the faces

back into the start position just by turning layers. When reassembling the cube it is, therefore, best to reconstruct the start position.

Warning 2 (moral): Cubologists who take their cube apart for other reasons than for mere curiosity or for lubrication (cf. section 1.3) forfeit their professional honor.

In order to facilitate the description of the motion of individual unit cubes we introduce simple abbreviations for the 6 faces, 12 edges and 8 corners of the cube. **From now on we always assume to have the cube in front of us in such a way that front, back, right, left, up and down can be clearly identified.** Using the small initials of these adjectives we can then identify the six faces of the cube: f = front, b = back, etc. (figure 3). Every edge is identified by the initials of the two adjoining faces in any order, every corner by the initials of the three adjoining faces, choosing the first face arbitrarily and continuing clockwise (looking at the corner from outside). We then have for example *ru* (or *ur*) for the right upper edge, *ubr* (or *bru* or *rub*) for the "up back right" corner, and *fdl* (or *dlf* or *lfd*) for the "front down left" corner (figure 4).

In every face, on every edge and at every corner there sits exactly one face, edge or corner cubie, respectively. Our abbreviations can,

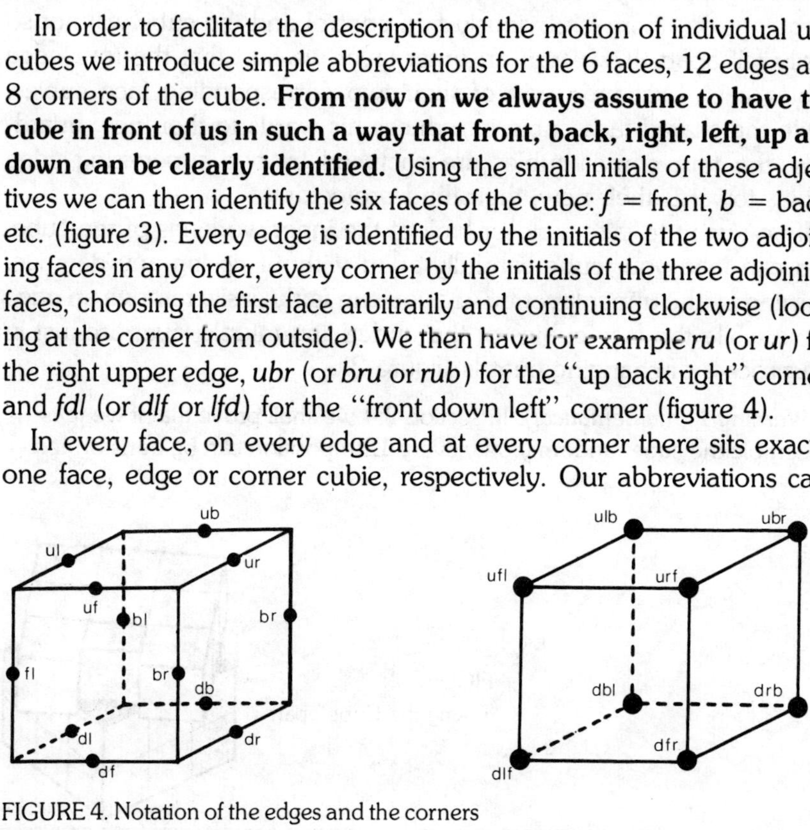

FIGURE 4. Notation of the edges and the corners

therefore, be used to identify the location–the "cubicle"–of a cubie (cf. also lat. cubare = lie, sleep). We thus speak about the edge cubie in cubicle *ru* (in short: edge cubie *ru*) or the corner cubie in cubicle *ubr* (in short: corner cubie *ubr*).

1.2 Maneuvers

By a *possible position* we mean a color pattern that can be arrived at from the start position by turning the 9 layers, and where two color patterns are considered equivalent if one can be turned into the other through a motion of the cube as a rigid body. In order to describe how any possible position can be returned to the start position we need a simple but sufficiently general notation for all the manipulations the cube permits.

The elementary manipulations are the moves. We distinguish *outer layer moves, middle layer moves* and *cube moves.*

An *outer layer move* is the turning of one of the 6 outer layers by 90° or 180°. To characterize the moves by means of colors seems impractical, not only because of the different coloring of the cubes available today, but also for a variety of other reasons. (Even if we prescribe the 6 colors there are still $5 \cdot 3 \cdot 2 = 30$ essentially different ways to color a cube: after choosing one color for one face there are 5 possibilities left for the opposite face; after choosing any of the 4 remaining colors for one of the remaining blanks we have 3 possibilities for the opposite face and finally 2 possibilities for the remaining 2 faces.) Instead let us assume as in the last section that the cube is laying in front of us in such a way that front, back, right, left, up and down are clearly defined. We then denote by the corresponding capitals

$$F, B, R, L, U, D$$

a 90°-turn of the corresponding layer, moving it clockwise, looking

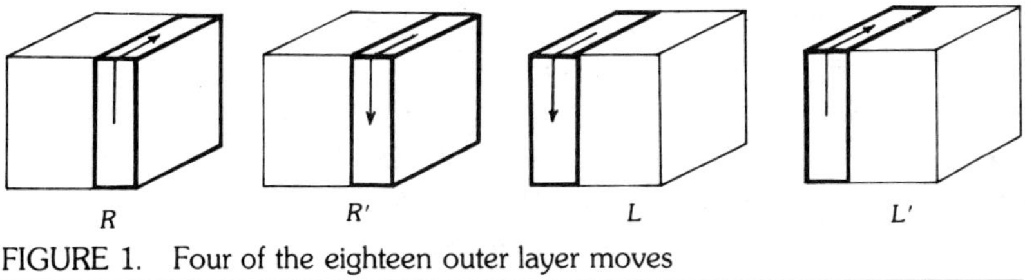

FIGURE 1. Four of the eighteen outer layer moves

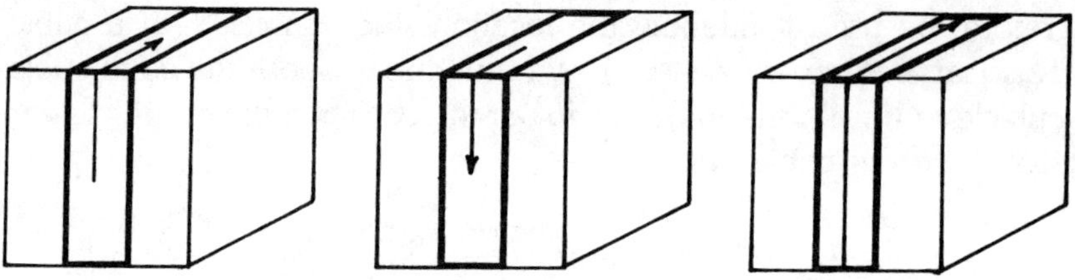

$M_R = M'_L$ $\qquad$ $M'_R = M_L$ $\qquad$ $M^2_R = M^2_L$

FIGURE 2. Three of the nine middle layer moves

from the outside at the layer in question. If the 90°-turn is counterclockwise we change the symbols to

$$F', B', R', L', U', D'$$

(figure 1). In the case of 180°-turns we add the exponent 2:

$$F^2, B^2, R^2, L^2, U^2, D^2.$$

To simplify matters, clockwise will from now on always be called *right,* and counterclockwise *left,* as is normal with screws.

On principle we could make do with outer layer moves only; such a restriction would also help to simplify the mathematical model. However, it is more convenient and above all agrees with the construction of the cube to also permit middle layer moves. A *middle layer move* is a 90° or 180° turn of one of the 3 middle layers. If *instead of an outer layer the adjoining middle layer* should be turned we simply put the letter M in front of the subscript of the symbol of the corresponding outer layer move: M_F, M'_F, M^2_F etc. That way each of the middle layer moves has 2 symbols: $M_R = M'_L$, $M'_R = M_L$, $M^2_R = M^2_L$ etc. (figure 2), but this redundancy has no drawbacks. Outer and middle layer moves will be called *layer moves.*

A *cube move,* finally is (for the moment, it will later be somewhat generalized) the turning by 90° or 180° of the whole cube as a rigid body around one of the three axes connecting the centers of opposite faces of the cube. We put the letter C in front of the subscript of the symbol of an outer layer move, if *instead of the outer layer the whole cube* is to be turned: C_F, C'_F, C^2_F etc. Hence these moves also have two names: $C_R = C'_L$, $C'_R = C_L$, $C^2_R = C^2_L$ etc. (figure 3).

Generally, of course, not just one but a whole sequence of moves has to be performed. We write them down *from left to right in the order in which they should be performed,* f.i.

$$R^2M'_DR^2M_D$$

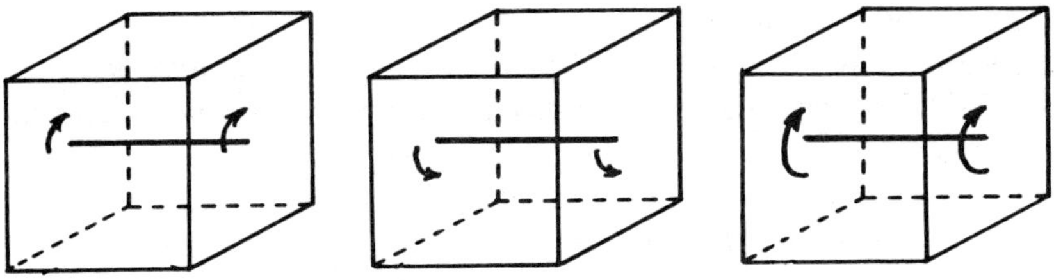

$C_R = C_L'$ $C_R' = C_L$ $C_R^2 = C_L^2$

FIGURE 3. Cube moves

means that we first have to perform the move R^2, then M_D', then again R^2 and finally M_D.

Caution: The order of the moves is important! *UR* for instance brings the edge cubie *ur* to *uf* whereas *RU* brings the same edge cubie *ur* to *br*.

Several consecutive moves of the same layer can always be condensed into one move, or they cancel each other out altogether; R^2R for example has the same effect as R', and $FRR'D$ the same as *FD*. This also applies to several consecutive cube moves around the same axis. For all practical purposes it is, therefore, no restriction if we give the following definition: An l-*maneuver is a finite sequence of moves in which neither two layer moves with the same layer nor two cube moves around the same axis follow each other.* (The letter l in l-maneuver refers to the "large model" of Section 2.3 where we consider not only outer layer moves but also middle layer moves and cube moves. We will, however, omit this letter l for reasons of simplicity if this is possible without causing confusion.)

One might argue that ruling out certain sequences of moves is really superfluous, especially since other sequences can also be condensed; for example RM_RL' has the same effect as C_R. However, this restriction will be useful in the next chapter, where we will form a *group* out of the *set of all maneuvers* as simply and as naturally as possible. We will then also see the deeper reason for obvious abbreviations already in use as for example $(R^2U^2)^3$ for $R^2U^2R^2U^2R^2U^2$, reminiscent of algebraic multiplication.

At the end of a maneuver we will generally indicate its length by giving the number of layer moves required in brackets. Cube moves, which–contrary to the layer moves–do not change the position (cf. the definition at the beginning of this section), are not counted. Example: $((M_RU)^4C_RC_U')^3$ (24). Incidentally, this maneuver is our favorite and we shall come across it several more times.

After this brief digression, we shall now answer another important question: how can we reverse a maneuver that has just been performed? This certainly poses no problem for a single move: the *inverse maneuver* of R is R', of R' is R, and R^2 is its own inverse etc. But how can RU be reversed? Right! The inverse maneuver for RU is not $R'U'$, but $U'R'$. In general, *the inverse maneuver m' of a maneuver m is arrived at by reading m backwards and inverting each single move.* After another simple example, $(R^2M_D'R^2M_D)' = M_D'R^2M_DR^2$, we give a somewhat more difficult one for more advanced cubologists: The inverse maneuver of $((M_RU)^4C_RC_U')^3$ (24) is $(C_UC_R'(U'M_R')^4)^3$ (24).

It isn't any different in our daily lives; if we can take individual actions back at all, we often have to begin with the last one. After all, if in the morning you first put on your underpants and then your jeans, in the evening you have no choice but to take your jeans off first. And a somewhat more morbit example: How do you get back to the start position when you are sitting in a car and 1. roll down the window and then 2. stick out your head to check the back tire?

1.3 A Simple Strategy

A *strategy* is a set of directions which describes precisely how to get back to the start position from any possible position of the cube. The *length* of a strategy is the number of layer moves required in the most unfavorable case. The shorter a strategy the more complicated it will generally be; i.e. it will be necessary to distinguish between many different cases, to use special maneuvers and perhaps even tricky advance calculations. An elegant compromise between brevity and simplicity is achieved through the following procedure, carried out in six stages.

Stage 1 (≤33 moves), the "opening", cleans up one–preferably the top layer (the whole layer–not just the face). It has proved helpful to start with an easily recognizable color, f.i. white. We begin by turning the cube in such a way that the white face cubie is facing upwards. Next we arrange the four "white corner cubies" (corner cubies that have one white tile) around the white face cubie. We do not pay *any* attention to the face cubies of the horizontal middle layer, all we have to do is make sure, that the *position of all the "white cubies" relative to each other* is correct. One move will always suffice to transport the first

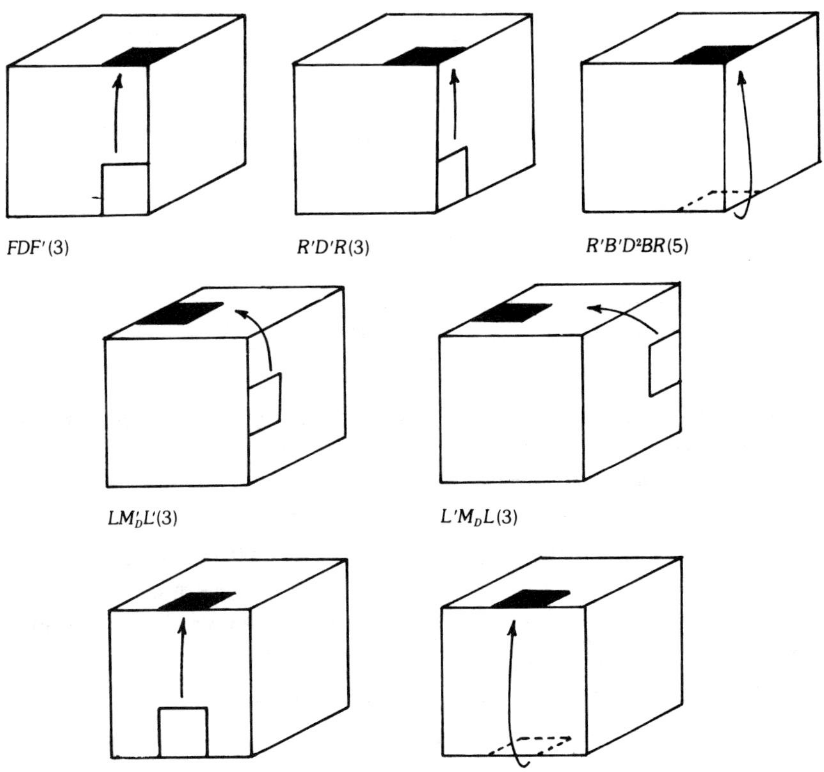

FIGURE 1. Maneuvers for the construction of the first layer

white corner cubie into the upper layer with its white tile facing upwards. Next, one of the two adjoining corner cubies is adjusted, then the two remaining corner cubies and finally the four "white edge cubies". Even if we refrain completely from thinking ahead, we need at most 1 (first corner cubie) +3 (second corner cubie) +9 (third and fourth corner cubie) +20 (the four edge cubies) = 33 moves to assemble the first layer. Figure 1 presents a collecting of maneuvers which transport a cubie from a cubicle outside the top layer into a prescribed cubicle in the top layer without causing any disorder in the rest of the top layer. By turning the whole cube about its vertical axis and moving the top layer each case can be reduced to one of the samples in figure 1. If all the white cubies are already in the upper layer, but not all of them lie in their proper cubicles or in their correct positions we simply use one of the maneuvers outlined above to pluck the troublemaker from the upper layer and then continue as indicated.

After the cube move C_R^2, the completed layer faces downward, and we proceed with **stage 2** (≤ 7 moves), which brings us to the "middle game". Again forgetting about the other layers, we begin by moving the four corner cubies of the new top layer into the correct cyclical arrangement, whereby it is irrelevant which of the three tiles of each corner cubie is facing upward. For four objects, 1, 2, 3, and 4, six cyclical arrangements are possible:

1	2	3	4	5	6
1 2	1 2	1 3	1 3	1 ↔ 4	1 ↔ 4
		↕ (3–2)	↕ (1–2)		
4 3	3 ↔ 4	4 2	2 4	3 2	2 ↔ 3

FIGURE 2. The six cyclical arrangements of four objects

If arrangement 1 is the correct one, then 2, 3, 4 and 5 merely require the swapping of two adjoining objects. Arrangement 6 demands a simultaneous swapping of the two front and the two back objects (cf. arrows in fig. 2).

How can we swap our corner cubies around without messing up the already completed bottom layer? Using the numbering given in the "maneuver index" at the end of this book, the two following maneuvers will do the trick:

$$m_{915} = RUR'U'F'U'F \ (7),$$
$$m_{333} = FRUR'U'F' \ (6).$$

Figures 3 and 4 demonstrate how these maneuvers are used.

Arrangements 6 and 3 can be straightened out without further ado

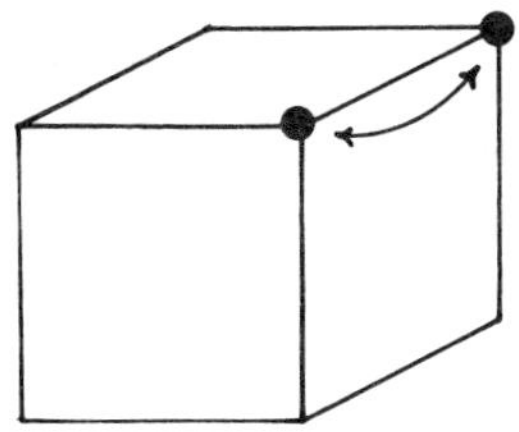

FIGURE 3.
The essential effect of
$m_{915} = RUR'U'F'U'F$(7)

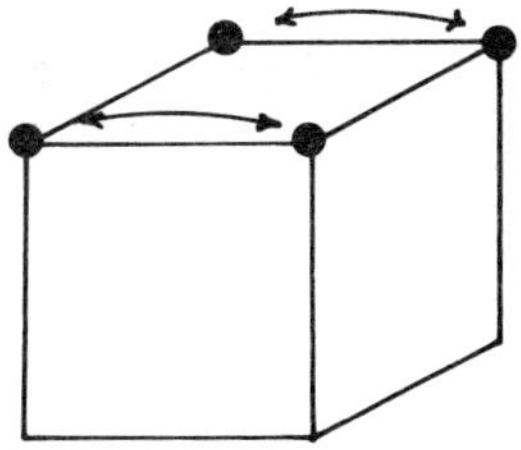

FIGURE 4.
The essential effect of
$m_{333} = FRUR'U'F'$(6)

by applying m_{333} and m_{915}, respectively. In the case of 2, 4 and 5 we first have to turn the cube around its vertical axis until the two corner cubies we wish to swap are on our right; we then proceed with m_{915}.

We recommend that the reader figure out a rule to determine as efficiently as possible–and without comparing the corner cubies of the top layer with those of the bottom layer–which of the 6 arrangements he is faced with. One hint: Each of the four upper corner cubies has–apart from the color in common with the face cubie of the upper layer–two "sub-colors". If two diagonally opposite corner cubies have one of these sub-colors in common, we are dealing with one of the arrangements 2 to 5. If this is the case for one pair of diagonally opposite corner cubies it is also true for the other. As much as possible one should forget about the bottom layer already completed until the remaining two horizontal layers are finished.

Stage 3 ($\leq$ 14 moves) provides for the correct orientation of the four upper corner cubies, i.e. we see to it that the correct color is facing upward. During this process we don't, of course, want to interfere with either the completed bottom layer or the cyclical arrangement of the upper corner cubies. Depending on whether a corner cubie requires a 120°-right-turn or a 120°-left-turn (fictitious turn around the space diagonal of the cube running through the corner cubie) we speak of a +120°-turn or a −120°-turn. According to the "first law of cubology" in Section 2.4 the sum of the angles of all required turns must be an integral multiple of 360°; meaning in this case 360°, −360°, or 0°. Consequently, apart from the case where there is nothing left to do, there are only four possible cases:

(a) three corner cubies require a right turn,
(b) three corner cubies require a left turn,

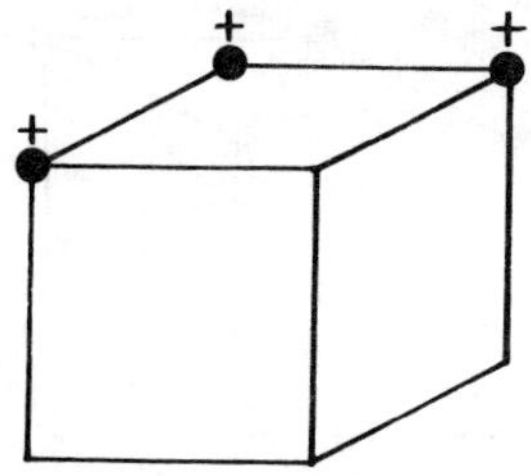

FIGURE 5.
The essential effect of
$m^*_{30} = R'U_2RUR'UR$(7)
(apart from U^2)

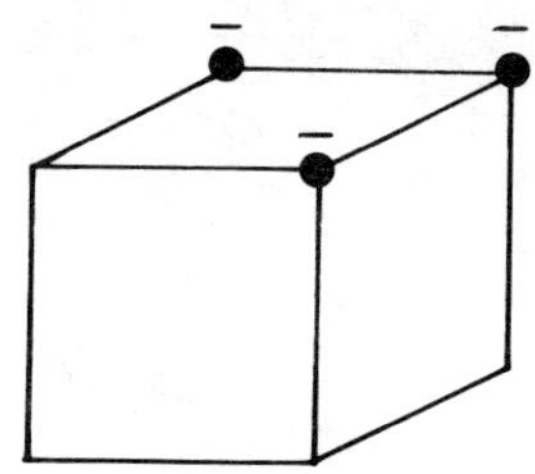

FIGURE 6.
The essential effect of
$m^{**}_{30}+ = LU^2L'U'LU'L'$(7)
(apart from U^2)

(c) one corner cubie requires a right, another one a left turn,
(d) two corner cubies require a right, the other two a left turn.

In case (a) we move the correctly oriented corner cubie to the right front corner of the cube by turning the cube around its vertical axis. We then apply

$$m^*_{30} = R'U^2RUR'UR \ (7)$$

and we are there (figure 5). In case (b) we follow the example of case (a), but do the left-right mirror image: We move the correctly oriented corner cubie to the left front corner of the cube and perform (left-handedly, if possible)

$$m^{**}_{30} = LU^2L'U'LU'L' \ (7)$$

(figure 6). Most people have a sufficiently developed sense of left-right symmetry to perform every maneuver which they know by heart also left-right reflected. (As an alternative we can also use the inverse maneuver $(m^*_{30})' = R'U'RU'R'U^2R$(7). Here the correctly oriented corner cubie must be moved to the left back corner.)

In case (c) the cube is turned around its vertical axis until the corner cubie requiring a left turn is sitting in the right front corner; we next apply m^*_{30} and find ourselves faced with case (b).

In case (d) we proceed similarly: one of the two corner cubies needing a right turn is moved to the right front corner of the cube; we carry out m^*_{30} and end up with case (a).

For stage 3 learning one maneuver of 7 moves proved sufficient; **stage 4** ($\leq$44 moves), which completes the top layer, manages with even less—here one maneuver of 5 moves is enough. With its help any u-edge-cubie (upper edge cubie) still stuck in the horizontal middle

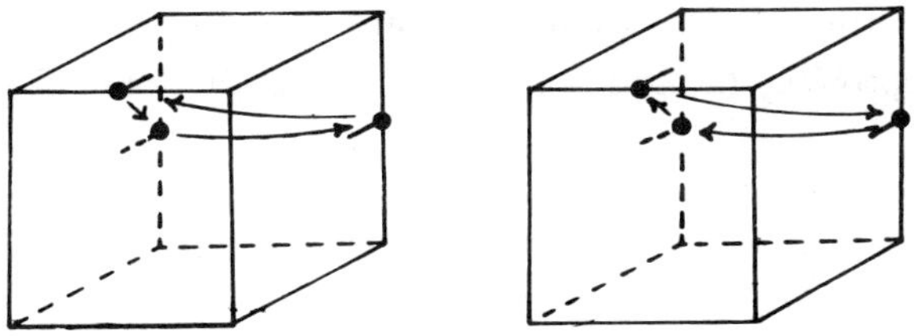

(a) $BM_RB^2M_R'B$(5) (b) $B'M_RB^2M_R'B'$(5)

FIGURE 7. The maneuvers of stage 4 of our simple strategy

layer can be speedily dispatched to its place in the upper layer—correct orientation included. This is done by turning the cube around its vertical axis until our edge cubie is sitting in the back of the horizontal middle layer with its prospective upper face looking either to the right or to the left but never backwards. If necessary, we next turn the upper layer until the home cubicle of our edge cubie is in the front of the cube. If the cubie is in the back on the right,

$$m_{520} = BM_RB^2M_R'B \ (5)$$

serves our purpose (figure 7a); in the case of "back left" the left-right reflected maneuver will do the task. It is, in this case, accidentally identical with the inverse of m_{520}:

$$m_{520}' = B'M_RB^2M_R'B' \ (5)$$

(figure 7b). If the four u-edge-cubies are already in the upper layer, but incorrectly placed, we first use m_{520} or m_{520}' to transport one of the displaced cubies back into the horizontal middle layer and then treat it as described.

With the completion of the top and the bottom layer we have reached our "end-game"- a mere pawn-pardon-edge cubie game. We enter **stage 5** ($\leq$6 moves) by first matching the colors of the three hori-

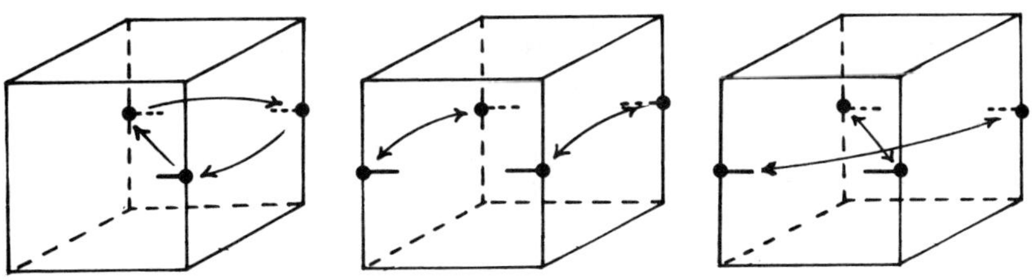

(a) $R^2M_D'R^2M_D$(4) (b) $(R^2M_D^2)^2$(4) (c) $(M_R^2M_D')^2$(4)

FIGURE 8. The maneuvers of stage 5 of our simple strategy

zontal layers (≤ 2 moves). At this point everything is in its place except possibly the four edge cubies of the horizontal middle layer. All these "straying" edge cubies can be sent back to their home cubicles in one stroke, by one of the three 4-move-maneuvers shown in figure 8. A preparatory turn of the cube as a rigid body may be necessary. The fact that one of these three maneuvers will always do the trick is, once again, a consequence of the first law of cubology in Section 2.4.

And it isn't quite by chance that we find ourselves again referring to this theorem, for it also states that even after all the cubies have found their cubicles, there will always be an even number of edge cubies "standing on their heads". This means that in **stage 6** (≤ 12 moves) either zero (and that would be the end of our game), or two, or all four edge cubies must be flipped. This is always possible by means of one of the three maneuvers compiled in figure 9. (Instead of m_{405_a} we can also apply maneuver $m_{405_b} = M_D R M_D R M_D R^2 \cdot M'_D R M'_D R M'_D R^2$ (12) which is somewhat easier to learn. An extremely simple maneuver for the reorientation of two edge cubies is $(M_R U)^4 (M_R U')^4$ (16)).

Hurrah! After, at the most, 116 layer moves our cube is back in its ardently searched for pristine state. A shift of the two moves, which match the three horizontal layers and merely serve to give beginners a clearer idea of where they stand, from the beginning of stage 5 to the very end permits us to skip the last move of all the maneuvers of stages 5 and 6, which shortens the strategy to 114 layer moves. With the help of further special maneuvers (cf. maneuver index) it is comparatively easy to condense our strategy to under 80 moves; stage 4 alone can be reduced by twenty moves. But even the unreduced simple strategy permits one to achieve "decent times". Applying the simple strategy, the author presently clocks in at an average of 50 seconds—not a bad time at all in view of his relatively old fingers; and he clearly owes it to the constant challenge of his colleague, Rolf Gall, who is always slightly

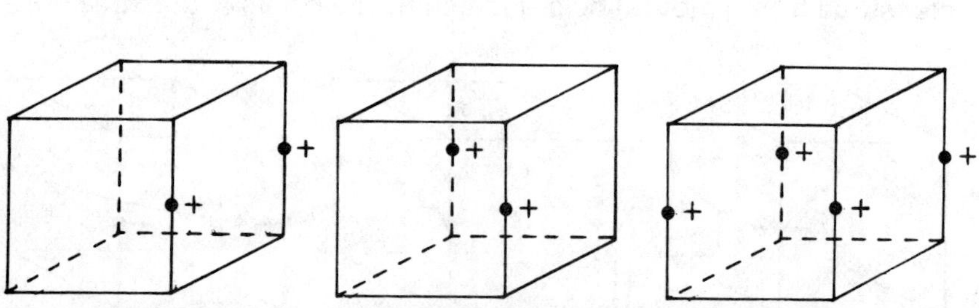

(a) $R'FD'RF' \cdot M'_D \cdot FR'DF'R \cdot M_D$(12)　(b) $R'FD'RF' \cdot M_D^2 \cdot FR'DF'R \cdot M_D^2$(12)　(c) $FD^2M'_RU^2FM_D^2 \cdot F'D^2M'_RU^2F'M_D^2$(12)

FIGURE 9. The maneuvers of stage 6 of our simple strategy

ahead and has in the meantime reached an average of 42 seconds. (It is understood that we "mix" each other's cube with special care and malice!) The record, however, is set at an average of 30 seconds; a feat which has four prerequisites: 1. A well oiled "racing cube"; vaseline, silicone oil or spray, or the ball bearing lubricant sold in bicycle stores all work well. 2. An economical strategy. This record is presently held by Morwan B. Thistlethwaite (London). Using a very large arsenal of maneuvers, partly discovered with the help of a computer program (cf. chapter 6), he needs at most 52 outer layer moves. 3. The ability to recognize on the spot which maneuver will get you one step further. Since only every twelfth out of all the conceivable positions is possible, the experienced cubologist usually needs only a quick glance at a small section of the cube. 4. A sleepwalker's certainty in the mastery of all the necessary maneuvers. Stored away in the brain, they must surface automatically and suddenly sit at the finger tips—like a sonata played by a gifted pianist.

1.4 RINIRENEFENEFI

The title of this section may go beyond the more or less extensive linguistic capacities of our readers. It evokes memories of the gibberish children enjoy so much. "You are dumb" in gibberish is "YITHEGOU

ITHEGARE DITHEGUMB". The answer is usually: "YITHEGOU ITHEGARE TITHEGOO." However, this is too involved for our purpose. RINIRENEFENEFI is an important word in the Rubikian language, a universal and pleasantly musical lingo, giving every cubologist the opportunity to express his essential requirements by means of only 8 consonants and 3 vowels. Even though we shall later return to the previously introduced more formalistic notation, in this section we want to oblige the more literary of our readers with a rapid course in Rubikian and a translation of the most important texts of our "simple strategy" into that language. In figure 1, all the components of that language are already contained:

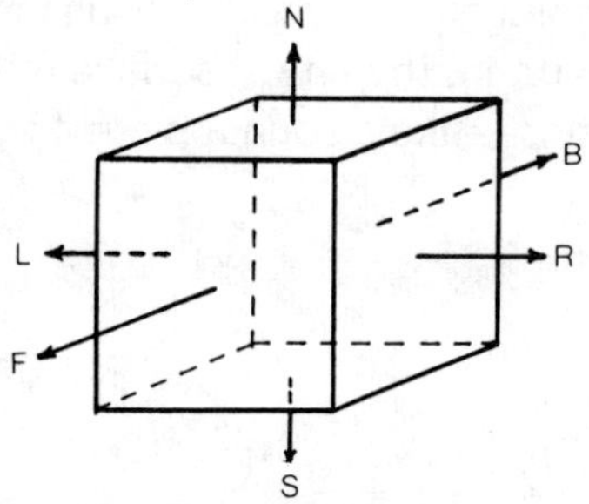

I = 90° right turn
E = 90° left turn
U = 180° turn (U-turn!)
M = middle layer
C = cube move

FIGURE 1. The basic Rubikian vocabulary

Only two letters are not compatible with the notation used so far: Rubikian, a very sound-conscious language, requires consonants at the beginning of each syllable and therefore calls the upper layer N, as in north, and (for reasons of symmetry) the bottom layer S, as in south. Each of the six consonants denoting the faces of the cube: R (Right), L (Left), F (Front), B (Back), N (North = up) and S (South = down) is followed by one of the three vowels I, E, or U. I denotes a 90°-right-turn of the corresponding layer, E signifies a 90°-left-turn and U a 180°-turn, a U-turn. The letter M or C after any of these vowels means that we have to turn the neighboring middle layer (M) or the whole cube (C, cube move) instead of the outer layer. To give an example: FI signifies a 90°-right-turn of the front layer. RUM turns the L-R middle layer by 180°; for left handers the Rubikian synonym LUM may here be more convenient.

RINIRENEFENEFI
Swaps two corners *right* away.
(figure 1.3.3)

FIRINIRENEFE
Two corner pairs swaps only she,
the magic goddess of cubology.
(figure 1.3.4)

Turn left three corners—one, two, three:
LINULENELINELE (figure 1.3.6).
Now turn them back the other way:
RENURINIRENIRI (figure 1.3.5).

Want to cycle three edges?
Don't ask a guy who hedges.
Ask Winnie the Pooh
and he'll tell you:
SEMRU SIMRU.
(Inverse of figure 1.3.8a)

RUSUM-RUSUM is really bright,
it swaps the edges left and right.
(figure 1.3.8b)

If all the edges are a mess
RUMSEM-RUMSEM will make it less.
(figure 1.3.8c)

The remaining "strategy-texts" are listed without mnemonic aids—a modest challenge to the reader:

BIRIM-BUREM-BI (figure 1.3.7a)
BERIM-BUREM-BE (figure 1.3.7b)
FISUREM-NUFISUM-FESUREM-NUFESUM (figure 1.3.9a)
REFISE-RIFESEM-FIRESI-FERISIM (figure 1.3.9b)
REFISE-RIFESUM-FIRESI-FERISUM (figure 1.3.9c)

While the last three texts are best remembered in waltz time, the first two ask for a somewhat more forceful accompaniment, for instance: Old McDonald had a farm, BIRIM-BUREM-BI! And on his farm he had some chicks, BERIM-BUREM-BE!

1.5 Operations

By now we can turn the cube back to its start position any time. It is, therefore, not much trouble to investigate the effect of individual maneuvers. This effect consists in certain cubies changing their positions. Corner cubies always move into corner cubicles, edge cubies into edge cubicles and face cubies into face cubicles. What we are dealing with are obviously house swapping rings among corner cubies, among edge cubies and among face cubies, respectively. (Occasionally some cubies are satisfied with merely installing themselves differently in their old cubicles, see f.i. figure 1.3.9.) Such a *change of position of cubies* which can be brought about by a maneuver is from now on called a *possible l-operation*. Again, the letter l in l-operation refers to the "large model" of section 2.3 and will be omitted if no confusion arises. To simplify matters further we also often omit the adjective "possible".

We start with a simple example: the 4-move maneuver m_{500} =

$R^2M_D'R^2M_D$ which we have already met in section 1.3. It produces an *edge-3-cycle* (see figure 1.3.8a), i.e. three edge cubies exchange their respective cubicles cyclically: edge cubie *fr* moves to *bl*, edge cubie *bl* to *br*, and edge cubie *br* into the cubicle *fr*. We write

$$R^2M_D'R^2M_D(4) \rightarrow (fr,\ bl,\ br),$$

and read "$R^2M_D'R^2M_D$ (4) brings about (*fr, bl, br*)". Instead of (*fr, bl, br*) we can also write (*bl, br, fr*), (*br, fr, bl*), (*rf, lb, rb*), (*lb, rb, rf*) or (*rb, rf, lb*); not however f.i. (*fr, lb, br*). For the notation for our operations should not only show which cubie moves into which cubicle, but also how the cubie settles down in it. (*fr, bl, . . .*) means that the edge cubie coming from *fr* settles in *bl* in such a way that its former *f*-facelet now lies in *b*, and its former *r*-facelet in *l*. More generally, for the notation of any *n*-cycle, i.e. of any house swapping ring of *n* cubies, we make the following agreement: (1) *After the opening parenthesis we write down one of the names of one of the cubicles involved in the house-swapping ring.* (Remember section 1.1: Each corner cubicle has three names, each edge cubicle two names, each face cubicle one name.) (2) *Next, the names of all the other cubicles that are part of the house swapping ring, are written down in the order determined by the sequence of the house swappings. The first letter always denotes that face of the cube into which the "first facelet" of the new cubie will fit, i.e. the facelet which was already situated in the old cubicle in the face mentioned first.* Automatically, the corresponding relation then holds for the second letter of all the names in case of an edge cycle and for the second and third letter of all the names in case of a corner cycle. (In the last case the regulation about a clockwise direction is essential.) At this point we urgently need some examples!

Figure 1 shows four different possible operations: the previously mentioned 3-cycle and three other ones, where the *same* edge cubies swap their cubicles in the *same* cyclical order of succession. Underneath each operation (whose graphic representation is undoubtedly self-explanatory) we write a maneuver which results in this operation. The complete notations for these edge-3-cycles can be found in the "maneuver-index" under m_{500}-m_{503}.

It follows from the "second law of cubology" in section 2.4 that for any 3 given edge cubies with a given cyclical order of succession, there are always precisely $2^2 = 4$ different "isolated" 3-cycles possible; "isolated" meaning here that the cube remains otherwise unchanged. For 3 corner cubies with given cyclical order of succession we even have $3^2 = 9$ isolated 3-cycles. Figure 2 provides an example. The complete

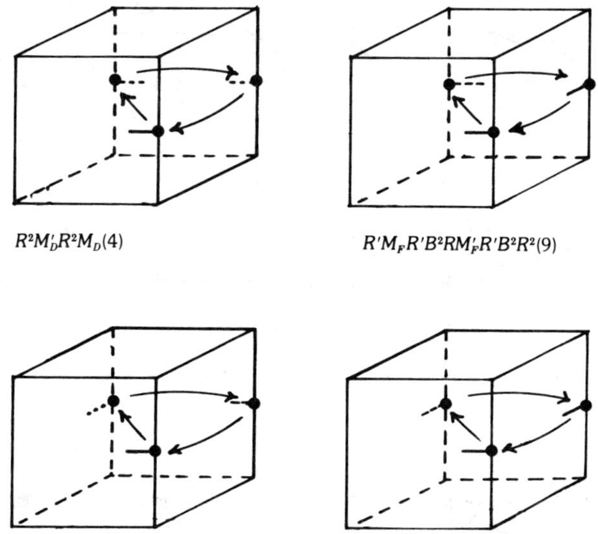

$R^2M_D'R^2M_D(4)$ $\quad$ $R'M_FR'B^2RM_F'R'B^2R^2(9)$

$B'M_RB \cdot R^2 \cdot B'M_R'B \cdot R^2(8)$ $\quad$ $B^2 \cdot RM_FR' \cdot B^2 \cdot RM_F'R'(8)$

FIGURE 1. Four possible isolated 3-cycles with the same edge cubies in the same cyclical order of succession

notation for these corner-3-cycles are to be found in the "maneuver-index" under m_{100}-m_{108}.

However, a maneuver generally sets not only one, but several cycles into motion, which are simply written down side by side. A few easy examples:

$$U \rightarrow (uf, ul, ub, ur)\ (ufl, ulb, ubr, urf),$$
$$M_R \rightarrow (f, u, b, d)\ (fu, ub, bd, df).$$

Figure 3 gives a few examples that are less trivial. They all show operations consisting of two 2-cycles ("transpositions") with "strictly fixed orientation" (meaning here: up remains up).

In one essential point, our notation for operations is still incomplete. What happens to the cubie sitting at the very end of the cycle-parenthesis? It will, of course, end up in the cubicle mentioned at the very beginning of the parenthesis. But how? It must, of course—be it corner or edge cubie—have the same rights as all its siblings to establish itself inside its new cubicle in two, respectively three different ways. This becomes immediately apparent in our maneuver $m_{333} = FRUR'U'F'$ (6), the effects of which are only roughly sketched in figure 1.3.4. This ma-

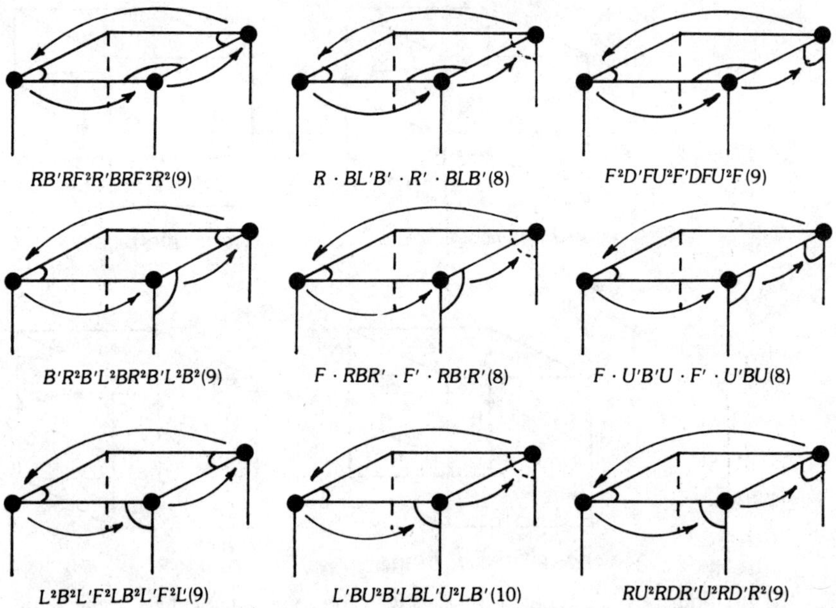

FIGURE 2. Nine possible isolated 3-cycles with the same corner cubies in the same cyclical order of succession

neuver sends the corner cubie from *urf* to *ufl* (up remains up) while the corner cubie from *ufl* is turned over to the right moving into its new cubicle *urf*. We write this down as (+*urf*, *ufl*) and introduce the following third point into our notation: (3) *Consider the cubie from the last cubicle within the cycle parenthesis moved—at least for the time being—into the first cubicle, assuming point (2) of our convention to be also valid for relationships between last and first cycle-elements. If the cubie must then be turned around to correspond with the given operation we write the sign + for a right turn of a corner cubie and the sign − for a left turn of a corner cubie, while an edge cubie requiring a turn is written with the sign + in front of the first cubicle in question, i.e. immediately after opening the parenthesis.* We obviously need more examples! Two of them are shown in figure 4, where we find, among others, the special case of a 1-cycle, which means a cubie returns to its old cubicle but establishes itself there in a new way.

It is important to distinguish carefully between the two concepts of the "maneuver" and the "possible operation". Already in section 1.3 we have met with two quite different maneuvers producing the same operation:

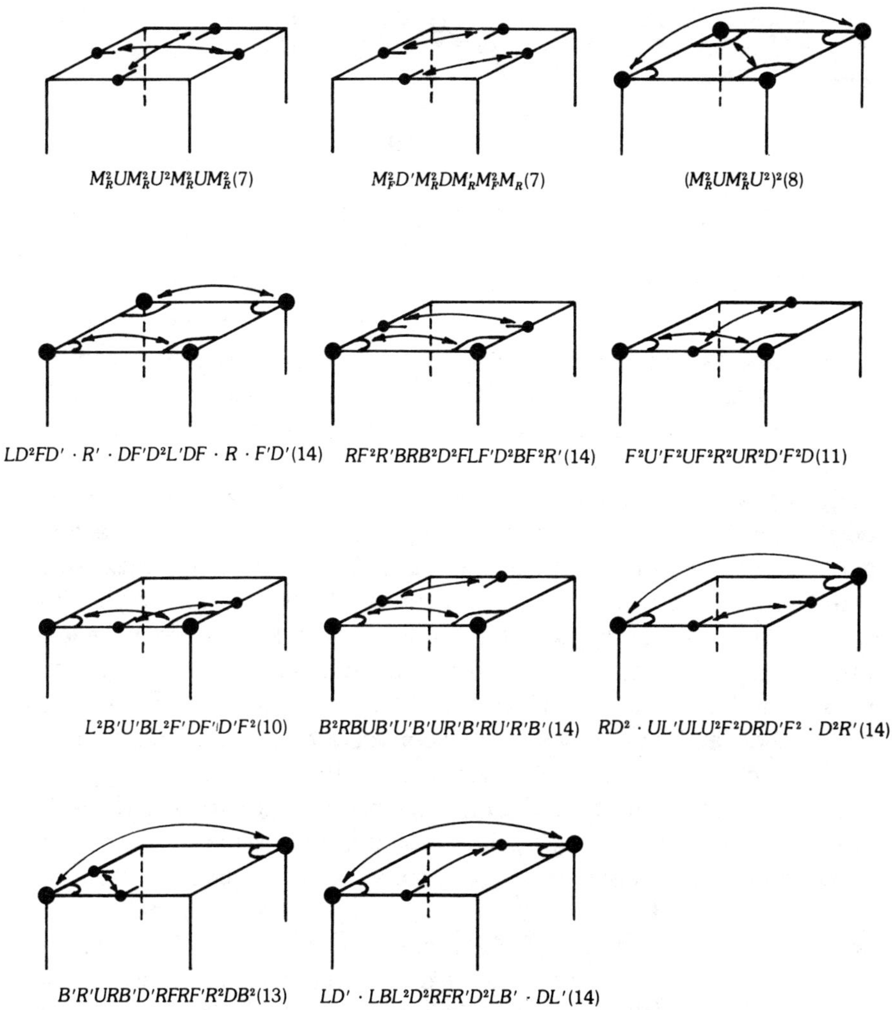

FIGURE 3. Eleven useful pairs of double swaps

$$m_{405} = R'FD'RF' \cdot M_D' \cdot FR'DF'R \cdot M_D \; (12) \rightarrow (+fr)\;(+br),$$
$$m_{405a} = M_DRM_DRM_DR^2 \cdot M_D'RM_D'RM_D'R^2 \; (12) \rightarrow (+fr)\;(+br).$$

Obviously, every possible operation can be brought about by infinitely many different maneuvers.

We shall next slightly extend our "maneuver" concept, as we already announced in section 1.2. This extension has to do with cube moves, moves which do not change the position but rather turn the

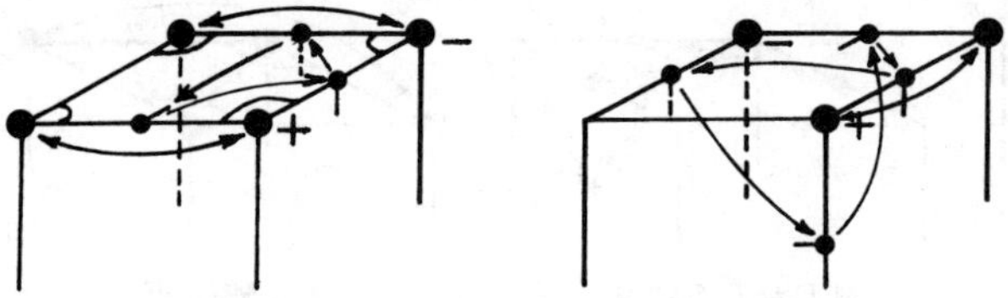

$m_{333} = FRUR'U'F'(6) \rightarrow$
$\rightarrow (+urf, ufl)(-ubr, ulb)$
(uf, ru, bu)

$m_{915} = RUR'U'F'U^{1}F(7) \rightarrow$
$\rightarrow (+urf, ubr)(-ulb)$
(ur, ul, rf, bu)

FIGURE 4. The precise effect of the maneuvers of stage 2 in our simple strategy

whole cube as a rigid body. One look can tell us that the cube can be brought back onto itself not only by turning it around its 3 "face axes", i.e. the axes connecting the centers of opposite faces, but also by two other kinds of turns around symmetry axes: by 180°-turns around "edge-axes", i.e. axes connecting the centers of diametrically opposed edges, and by 120°-turns around "corner axes"; i.e. axes connecting two diametrically opposed corners (figure 5). There are 3 face axes with 3 turns each, 6 edge axes with 1 turn each, and 4 corner axes with 2 turns each, amounting to a total of 23 different turns around axes. Even though the effect of a turn around an edge or corner axis can also be achieved by means of two consecutive turns around face axes, it is not justified and sometimes even inconvenient to prefer the 9 turns around face axes over the 14 other possible turns. Consequently, *from now on we shall count the 6 turns around the edge axes and the 8 turns around the corner axes among the cube moves and allow them in maneuvers.* In keeping with the notation used for turns around face axes we denote a turn around an edge or a corner axis by the letter C and two or three subscript capitals, corresponding to the notation of one of the two edges or corners lying on the rotation axis. In the case of a corner axis turn, the direction of the turn is determined by the order in which the three letters are written. Examples:

$$C_{UF} = C_{FU} = C_{DB} = C_{BD}$$

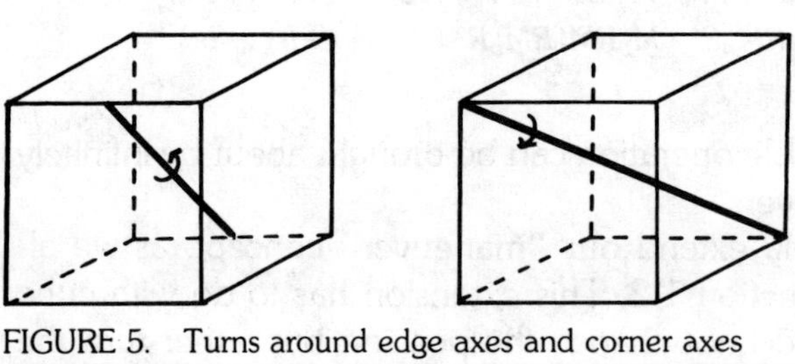

FIGURE 5. Turns around edge axes and corner axes

are the four names for the 180°-turn around the edge-axis connecting the centers of the edges *uf* and *db*.

$$C_{UFL} = C_{FLU} = C_{LUF} = C_{DBR} = C_{BRD} = C_{RDB}$$

are the six names for the 120°-turn around the corner axis *ufl-drb,* which moves face *u* to *f*, face *f* to *l*, and face *l* to *u*. The inverse turn to this would be C_{LFU} or C_{ULF} etc.

As a first application, we replace our favorite maneuver (cf. section 1.2) by a slightly more elegant one—which will be our new favorite, and write down the operation belonging to it:

m_{490} = $((M_R U)^4 C_{ULF})^3$ (24) →
(+*uf*) (+*ul*) (+*ub*) (+*ur*) (+*df*) (+*dl*) (+*db*) (+*dr*) (+*fl*) (+*lb*) (+*br*) (+*rf*).

The precise description of the effects of a cube move written down in our notation for operations is both a little exercise and a preparation of speculations we will indulge in later. Here we choose one for each of the different types of cube moves:

C_U → (*f, l, b, r*) (*uf, ul, ub, ur*) (*df, dl, db, dr*) (*fl, lb, br, rf*) (*ufl, ulb, ubr, urf*) (*dlf, dbl, drb, dfr*).

C_U^2 → (*f, b*) (*l, r*) (*uf, ub*) (*ul, ur*) (*df, db*) (*dl, dr*) (*fl, br*) (*fr, bl*) (*ufl, ubr*) (*ulb, urf*) (*dlf, drb*) (*dbl, dfr*).

C_{UF} → (*u, f*) (*d, b*) (*l, r*) (+*uf*) (+*db*) (*ub, fd*) (*ul, fr*) (*ur, fl*) (*dl, br*) (*dr, bl*) (*ufl, fur*) (*dbl, bdr*) (*ulb, frd*) (*ubr, fdl*).

C_{UFL} → (*u, f, l*) (*d, b, r*) (*uf, fl, lu*) (*db, br, rd*) (*ur, fd, lb*) (*dl, bu, rf*) (+*ufl*) (−*drb*) (*urf, fdl, lbu*) (*dfr, bld, rub*).

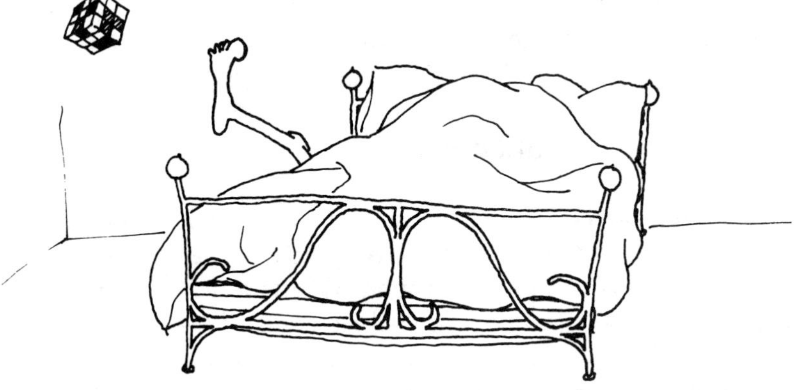

2 THE MATHEMATICAL MODEL

Chapter 2 and the first section of chapter 3 contain some mathematics. If this seems too unfamiliar, too hard, or too much burdened by unpleasant school memories, one can just skim over this part, murmuring "blah-blah" at every formula, and still understand the subsequent chapters.

However, from the close relationship between cube and mathematics, which one does not right away expect from dealing naively with the little object, the mathematician as well as the sporty cubist racing against the clock can profit. On one hand the cube illustrates a wealth of definitions and theorems, mainly from group theory, in a simple and attractive manner, where previously only trivial or much too complicated and abstract examples were available. On the other hand, we need a theory for a really satisfactory *understanding of the cube,* which includes answers to questions like: Which (and not only how many) positions can we reach from the start position by turning the layers? How are we restricted if only a part of all the moves is permitted? Is there any non-trivial operation for which it makes no difference whether we perform it before or after another operation?

In the first two sections the necessary basic concepts are assembled.

2.1 Sets and Functions

We assume that the reader has some notion of the naive concept of the *set*. According to Georg Cantor (1845–1918), the founder of the set theory, "a set A is an accumulation of certain distinct objects of our vision or our thought–the so-called elements of A–as a whole". The expression "a is an element of A" is abbreviated like this: $a \in A$ or $A \ni a$. A set is defined by an enumeration or a suggested enumeration of its elements in brackets, or by indicating one quality which all the elements of the set have in common, but which no other objects of our vision or our thought possess. Examples:

$\{0, 1, 2\}$ is the set of the three numbers 0, 1 and 2.

$\mathbb{N} := \{1, 2, 3, \ldots\}$ is the set of the "natural numbers" occurring when counting naturally.

$\mathbb{Z} := \{\ldots, -2, -1, 0, 1, 2, \ldots\}$ is the set of integers.

$P :=$ set of all possible positions for Rubik's magic cube
$= \{p \mid p$ is a possible position for Rubik's magic cube$\}$.

The last set plays a principle part in this chapter. A colon on one side of the equal-sign means that the symbol on this side is new and will be defined by the term on the other side of the equal-sign.

A set A is called a *subset* of a set B (notation: $A \subseteq B$ or $B \supseteq A$) if each element of A is also an element of B. For instance $\mathbb{N} \subseteq \mathbb{Z}$ holds true. There is one set which is a subset of every set: the *empty set* denoted by $\varnothing$ which contains no element.

Assume A and B to be arbitrary sets; then $A \cup B$ denotes their *union*, i.e. the set of elements contained in A or in B (or in both of them). $A \cap B$ denotes their *intersection*, i.e. the set of elements lying in A as well as in B. More generally, $A_1 \cup \ldots \cup A_n$ or $A_1 \cap \ldots \cap A_n$ stands for the union or the intersection of n sets $A_1, \ldots, A_n$ (where $n \in \mathbb{N}$). A and B are called *disjoint*, if they have no elements in common; in other words, if $A \cap B = \varnothing$.

A *function* f of the set A into the set B is a "rule" which assigns to each element $a \in A$, which we call the argument, precisely one element $b \in B$, which we call the image element, in symbols:

$$f\colon A \to B \text{ and } a \to af = b.$$

Notice that in this book the symbol for a function is on principle written on the right side of the argument.

As examples of functions we mention the rule which assigns to each natural number its square (f.i. $3 \to 9$), the rule which assigns to each maneuver the operation which the maneuver brings about (f.i. $R^2M_D'R^2M_D$ (4) $\to$ (fr, bl, br)), and the rule which assigns to each human being his father (f.i. Ernő Rubik jun. → Ernő Rubik sen.). (By the way, since only finitely many human beings have lived so far, not every human being has had a human being as his father.)

A function $f\colon A \to B$ is called *injective*, if the image elements of different arguments are always different. If every element of B appears as an image element, the function $f\colon A \to B$ is called *surjective*. A function which is injective as well as surjective is called *bijective* (or a *bijection*).

If $f: A \to B$ is bijective, for every $b \in B$ there is exactly one $a \in A$, such that $af = b$; i.e. we have a natural rule to assign to every $b \in B$ an $a \in A$. This function is called the inverse function of f and will be denoted by f'. (Professional mathematicians prefer the notation f^{-1}).

If $f: A \to B$ and $g: B \to C$ are two functions, we denote the function from A to C, which assigns to each $a \in A$ the element $(af)g$, by $f \circ g$:

$$f \circ g: A \to C, \quad a(f \circ g) = (af)g.$$

$f \circ g$ is called the *composition* of the two functions f and g. **Notice, that in this book functions that are to be applied one after the other are always written from left to right in the order of their application.** If $h: C \to D$ is a third function, we have for every $a \in A$ the equations

$$a((f \circ g) \circ h) = (a(f \circ g))h = ((af)g)h = (af)(g \circ h) = a(f \circ (g \circ h)),$$

and this means

$$(f \circ g) \circ h = f \circ (g \circ h).$$

Let A_1 and A_2 be two sets; then $A_1 \times A_2$ denotes the set of all ordered pairs (a_1, a_2) with $a_1 \in A_1$ and $a_2 \in A_2$:

$$A_1 \times A_2 := \{(a_1, a_2) | \ a_1 \in A_1 \text{ and } a_2 \in A_2\}.$$

Here we have by definition $(a_1, a_2) = (b_1, b_2)$ if and only if both $a_1 = b_1$ and $a_2 = b_2$. The set $A_1 \times A_2$ is called the *Cartesian product* of the sets A_1 and A_2 (in this order). If $|A|$ is in general the number of

elements of a finite set A, we have for two finite sets A_1 and A_2 the formula

$$|A_1 \times A_2| = |A_1| \cdot |A_2|.$$

The Cartesian product of n sets $A_1, \ldots, A_n$ ($n \in \mathbb{N}$) is defined analogously:

$$A_1 \times \ldots \times A_n := \{(a_1, \ldots, a_n) | a_1 \in A_1, \ldots, a_n \in A_n\}.$$

The elements of this set, the "n-tuples" $a = (a_1, \ldots, a_n)$, can be thought of as functions of the n-element set $\{1, \ldots, n\}$ into the set $A_1 \cup \ldots \cup A_n$ with $ia \in A_i$ for $i = 1, \ldots, n$, were a_i stands for ia, that is for the image element of the argument i under the function a. The formula mentioned above for the number of elements of the Cartesian product of two finite sets A_1 and A_2 holds similarly for n finite sets ($n \in \mathbb{N}$):

$$|A_1 \times \ldots \times A_n| = |A_1| \cdot \ldots \cdot |A_n|.$$

In the case $A_1 = A_2 = \ldots = A_n = A$ we write A^n instead of $A_1 \times \ldots \times A_n$, and the last formula simply becomes

$$|A^n| = |A|^n.$$

Now we come to the last concept of this section. In many sets a "relation"–a kind of kinship between individual elements–is defined in a natural manner: An integer a does or does not differ from an integer b by an integral multiple of 3, a position a of Rubik's cube can or cannot be transformed into a position b in at most one move, a maneuver a can or cannot cause the same operation as a maneuver b. We write $a \sim b$, if the element a is related to the element b.

Definition. A relation $\sim$ in the set A is called an *equivalence relation,* if the three following conditions hold:

(R) $a \sim a$ for every $a \in A$ ("reflexivity"),

(S) $a \sim b$ implies $b \sim a$ ("symmetry"),

(T) $a \sim b$ and $b \sim c$ imply $a \sim c$ ("transitivity").

Among the three relations mentioned above, only the second one is not an equivalence relation, since it does not possess the property (T). We take a closer look at the first relation.

Example. For every $r \in \mathbb{N}$ the relation in the set $\mathbb{Z}$ of integers (or any subset of $\mathbb{Z}$, f.i. $\mathbb{N}$) defined by

$$a \sim b \text{ if and only if } a - b \text{ is an integral multiple of } r$$

is an equivalence relation: Since $a - a = 0 = 0 \cdot r$, we have (R). $a - b = k \cdot r$ implies $b - a = (-k) \cdot r$, hence (S) holds. $a - b = k_1 \cdot r$ and $b - c = k_2 \cdot r$ imply $a - c = (a - b) + (b - c) = (k_1 + k_2 + k_2) \cdot r$, hence (T) holds.—If $a - b$ is an integral multiple of r we often write

$$a = b \text{ modulo } r$$

or $a \equiv b$ (r) ("a is congruent to b modulo r") or similarly. From dealing with the time of day we have long been familiar with congruence calculations: 5 hours past 10 a.m. is 3 p.m., since $10 + 5 = 3$ modulo 12.

If $\sim$ denotes an equivalence relation in the set A and $a \in A$, the set $C_a := \{b \in A | \, b \sim a\}$ is called the *equivalence class* of a. Assume two equivalence classes have one element in common: $c \in C_a \cap C_b$. Then, if d is an arbitrary element of C_a, that is $d \sim a$, then since $a \sim c$ and $c \sim b$, it follows that $d \sim b$, that is $d \in C_b$, hence $C_a \subseteq C_b$; and since we can similarly show $C_b \subseteq C_a$, we get $C_a = C_b$. Two equivalence classes having one element in common are, therefore, always identical, or stated in another way, two distinct equivalence classes are always disjoint. If, on the other hand, A is divided up into any disjoint non-empty subsets, an equivalence relation is defined by

$$a \sim b \text{ if and only if } a \text{ and } b \text{ lie in the same subset.}$$

The definition of an equivalence relation in A and the decomposition of A into disjoint non-empty subsets is, therefore, one and the same thing. If $f\colon A \to B$ is an arbitrary function, then

$$a \sim b, \text{ if and only if } af = bf,$$

defines an equivalence relation in A. Every equivalence relation is of this type, since, given an equivalence relation, we can choose for f the function which assigns to each a its equivalence class C_a.

2.2 Groups

Mathematicians do not work with numbers only. In a natural manner they arrive at "operating on" number rows (vectors), number arrays (matrices), sets, functions and many other things; in other words, to an ordered pair of two of these items a third is assigned. To learn all the essential "calculating rules" for every single one of the various operations separately would overtax mathematicians even more severely than they already are. Luckily, very frequently the same rules apply, so that the situation remains under control, despite the great variety of things connected in such a manner. For every actual operation it is helpful to figure out a few "basic rules", from which all the other rules can be deduced. In the case of two different operations with possibly quite different items this allows us to merely compare these basic rules, to see whether all the deduced rules of the two systems agree. A very frequent and therefore important case is an operation fulfilling the basic rules (A), (E), (I) of the following definition. For simplicity's sake the operation is denoted by a multiplication sign · which we shall later mostly omit. However, it need not have anything to do with the ordinary multiplication of numbers.

Definition 1. The set A is called a *group,* if to any two elements a and b of A an element a · b of A is assigned, and the following three rules hold:

(A) $(a \cdot b) \cdot c = a \cdot (b \cdot c)$ for all $a, b, c \in A$ ("associativity").

(E) There exists an element $e \in A$ with $e \cdot a = a \cdot e = a$ for all $a \in A$ ("existence of a neutral element e").

(I) For every $a \in A$ there exists an $a' \in A$ with $a \cdot a' = a' \cdot a = e$ ("existence of inverse elements").

Not required, however, of a group is

(C) $a \cdot b = b \cdot a$ for all $a, b \in A$ ("commutativity").

If (C) holds for a group we talk about a *commutative* or *abelian* group (named after the Norwegian mathematician N. H. Abel, 1802–1829).

An immediate consequence of associativity is the so-called *generalized associative law:* Whenever we have a sequence of elements to be connected by an operation, any syntactically meaningful manner of bracketing the sequence leads to the same result; for example $a_1(((a_2a_3)a_4)a_5) = ((a_1a_2)(a_3a_4))a_5$, so that the brackets can be omitted altogether and we simply write $a_1a_2 \ldots a_n$. Moreover, the neutral element is uniquely defined: Assuming e_1 and e_2 to be two neutral elements, we have $e_1 = e_1e_2 = e_2$. Finally, an equation of the form $ax = b$ ($a, b \in A$ given, $x \in A$ required) has always exactly one solution: By multiplication from the left with a' it follows $a'(ax) = a'b$, hence $x = a'b$; in other words, only $x = a'b$ is possible as a solution, and $x = a'b$ is, due to $a(a'b) = (aa')b = eb = b$, a solution. Similarly, one sees that the equation $ya = b$ ($a, b \in A$ given, $y \in A$ required) has always exactly one solution, namely $y = ba'$. The special case $b = e$ shows that the inverse element of an element a is always uniquely determined, and this is a belated justification for the notation a'.

If a certain group lends itself more to ordinary addition than to multiplication we prefer + as an operating symbol, particularly with abelian groups. In these cases we denote the neutral element by 0 and the inverse element of an element a by $-a$. The equations in (A), (E) and (I) are then written as: $(a + b) + c = a + (b + c)$, $0 + a = a + 0 = a$, and $a + (-a) = (-a) + a = 0$.

Example 1. The set $\mathbb{Z} = \{\ldots, -2, -1, 0, 1, 2, \ldots\}$ of integers is an abelian group with respect to ordinary addition +: Here everybody is familiar with the equations just mentioned. With respect to ordinary multiplication $\cdot$, the set $\mathbb{Z}$ is not a group, since (I) is not fulfilled. On the other hand, the set $\{\ldots, -4, -2, 0, 2, 4, \ldots\}$ of all even integers is with the ordinary addition a group again. Compared to $\mathbb{Z}$ it provides us with an example for

Definition 2. A non-empty subset B of a group A is called a *subgroup* of A, if B is a group with the operation defined in A.

Trivial subgroups of every group A are A itself and the set containing only the neutral element. In general, a non-empty subset B of a group is apparently a subgroup, if it contains with every two elements a and b also ab, with every element a also a'. In the case of *finite groups* (groups with finitely many elements) we can even do without the last condition: a non-empty finite subset B of an arbitrary group A is a subgroup, if and only if it contains with every two elements a and b also ab. For if a is any element of B, all the elements of the form ab with $b \in B$ differ from each other (the equation $ax = c$

has always exactly one solution), hence every element of B, particularly e, must be one of the elements ab, and we have $a' \in B$. The number of elements of a finite group A is called the *order* $|A|$ of A. That for every $n \in \mathbb{N}$ there exists a group or order n is shown in

Example 2. Let $C_n := \{0, 1, \ldots, n-1\}$ and for $a, b \in C_n$

$$a \oplus b := \begin{cases} a + b, & \text{if } a + b < n \\ a + b - n, & \text{if } a + b \geq n, \end{cases}$$

in brief

$$a \oplus b \in C_n \text{ with } a \oplus b = a + b \text{ modulo } n.$$

With the operation $\oplus$ the set C_n is an abelian group, the so-called *cyclic group of order n*. If it is obvious that the operation in C_n is referred to, we simply omit the circle around the+ sign. In C_3 we then have, for instance, the equations $1 + 2 = 0$ and $2 + 2 = 1$.

In every operation (in the sense of section 1.5), corner, edge or face cubies can swap their cubicles. Hence, every operation gives in a natural way a bijective function of the 8-element set of all the corner cubies into itself, a bijective function of the 12-element set of all the edge cubies into itself and a bijective function of the 6-element set of all the face cubies into itself. Of particular importance for our theory is, therefore,

Example 3. Let S_A be the set of all bijective functions of a set A into itself. S_A is a group with composition as its operation: The associativity

$$(f \circ g) \circ h = f \circ (g \circ h) \quad (f, g, h \in S_A)$$

was checked already in the last section. The neutral element is the *identity function* denoted by I (or more precisely by I_{S_A}), defined by $aI = a$ for all $a \in A$. The inverse function of f, already previously denoted by f', is inverse to f also in the group S_A. S_A is called the *symmetric group* over A. Its elements are known as *permutations of A*. We shall take a closer look at the case $A = \{1, 2, \ldots, n\}$, where S_n is written instead of S_A.

Theorem 1. $|S_n| = 1 \cdot 2 \cdot \ldots \cdot n =: n!$ (read "n factorial").

Proof. To determine a permutation of $\{1, \ldots, n\}$ we have to assign to every element of the set $\{1, \ldots, n\}$ an element of the set $\{1, \ldots, n\}$ as an image element, keeping in mind that every element may appear only once as an image element. Proceeding one step at a time, i.e., first choosing the image element of 1, then the image element of 2 etc., we have n possibilities for 1, then for the image element of 2 there are only $n - 1$ possibilities left and so on, all in all $n \cdot (n-1) \cdot \ldots \cdot 1$ possibilities. □

Any arbitrary permutation can be represented as the product of disjoint *cycles*, the way it was already done in section 1.5 with corner, edge, and face cubies. We write

$$f = (i_1, i_2, \ldots, i_r)$$

and call f an *r-cycle,* if $i_1, i_2, \ldots, i_r$ are r different elements of $\{1, \ldots, n\}$, and $i_1f = i_2, i_2f = i_3, \ldots, i_rf = i_1$ as well as $if = i$ for all other $i \in \{1, \ldots, n\}$ hold. The $3! = 6$ elements of S_3, for instance, are:

$$S_3 = \{I, (1, 2), (1, 3), (2, 3), (1, 2, 3), (1, 3, 2)\}.$$

2-cycles, i.e. permutations swapping only two elements and keeping all the others fixed, are also called *transpositions,* and products of two disjoint transpositions are called *double transpositions.* (Two cycles $(i_1, \ldots, i_r)$ and $(j_1, \ldots, j_s)$ are called disjoint, if the sets $\{i_1, \ldots, i_r\}$ and $\{j_1, \ldots, j_s\}$ are disjoint.) From now on we write simply fg instead of $f \circ g$.

For $n \geq 3$ the symmetric group S_n is not abelian, since for instance $(1, 2)(2, 3) = (1, 3, 2)$, but $(2, 3)(1, 2) = (1, 2, 3)$. (According to our convention the transpositions are to be carried out from left to right.) However, S_n is even in an extreme way non-abelian. For this we need

Definition 3. Let A be any group. Then

$$Z(A) := \{a \in A |\, ab = ba \text{ for all } b \in A\}$$

is called the *center* of A.

$Z(A)$ is always a subgroup of A. ($a, b \in Z(A)$ implies $ab \in Z(A)$, since for any $c \in A$ always $abc = cab$. Similarly, $a \in Z(A)$ implies $a' \in Z(A)$, since for every $b \in A$ always $a'b = (b'a)' = (ab')' = ba'$.) Now we can prove

Theorem 2. For $n \geq 3$ the center $Z(S_n)$ of S_n contains only the neutral element I.

Proof. If $f \in S_n$ and $f \neq I$, we have an element $a \in \{1, \ldots, n\}$ with $af =: b \neq a$. Since $n \geq 3$ there exists an element $c \in \{1, \ldots, n\}$ differing from a and b. Then $af(a,c) = b(a, c) = b$ and $a(a, c)f = cf \neq af = b$, hence $f(a, c) \neq (a, c)f$. □

A pair (i, j) of elements of the set $\{1, \ldots, n\}$ is called an *inversion* of the permutation f, if $i < j$ and $if > jf$.

A permutation is called *even* or *odd,* respectively, if the number z of its inversions is even or odd, respectively. $(-1)^z$ is called the *sign* of the permutation f and is also denoted by sgn f ("signum of f"). In the group S_n, even and odd permutations behave like even and odd integers with addition: The product (composition) of two even or two odd permutations is even, the product of an even with an odd permutation is odd. This follows from

Theorem 3. $\text{sgn}(fg) = \text{sgn}\, f \cdot \text{sgn}\, g$.

Prove. We have

$$\text{sgn}\, f = \prod_{1 \leq i < j \leq n} \frac{jf - if}{j - i},$$

since in this product, which runs over all pairs (i, j) of elements from $\{1, \ldots, n\}$ with $i < j$, except for the sign, the numerator has exactly the same factors as the denominator, and there are as many negative factors as inversions. (For $k, l \in \{1, \ldots, n\}$ with $1 \leq k < l \leq n$, the numbers kf' and lf' are different; that is, there exist $i, j \in \{1, \ldots, n\}$ with $l \leq i < j \leq n$ such that either $jf = k$ and $if = l$ or $jf = l$ and $if = k$.) It follows

$$\operatorname{sgn}(fg) = \prod_{i<j} \frac{jfg - ifg}{j - i} = \prod_{1<j} \frac{jf - if}{j - i} \cdot \frac{jfg - ifg}{jf - if}$$

$$= \prod_{i<j} \frac{jf - if}{j - i} \cdot \prod_{1<j} \frac{jfg - ifg}{jf - if} = \operatorname{sgn} f \cdot \operatorname{sgn} g. \qquad \square$$

Because of Theorem 3 the set of all even permutations of S_n is a subgroup of S_n, the so-called *alternating group* A_n. For $n \geq 2$ there are as many even permutations as odd permutations ($f \rightarrow f(1, 2)$ is a bijection between even and odd permutations), i.e. the order of A_n is $n!/2$.

Theorem 4. For $n \geq 2$ every element of S_n is a product of (in general not disjoint) transpositions.

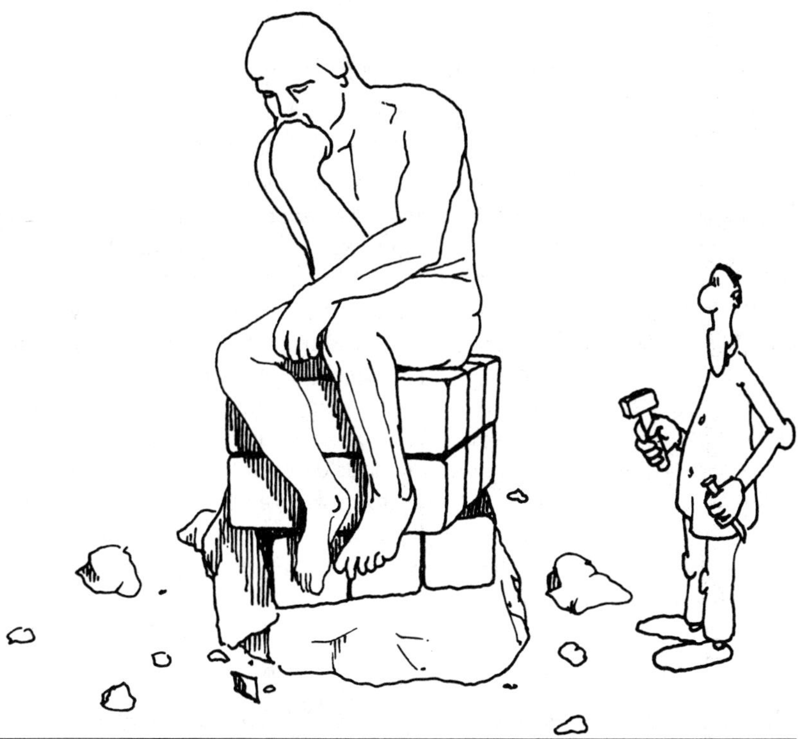

Proof. Since every permutation is the product of cycles, it suffices, to prove the claim for cycles. Indeed, $(i_1, \ldots, i_{r-1}, i_r) = (i_1, \ldots, i_{r-1})(i_1, i_r) = \ldots = (i_1, i_2)(i_1, i_3) \ldots (i_1, i_r)$, i.e. every r-cycle can be written as the product of $r - 1$ transpositions. □

Obviously, the intersection of arbitrary many subgroups of a group A is also a subgroup of A. Therefore, for every subset C of A there always exists a smallest subgroup B containing all the elements of C, namely the intersection of all subgroups of A which contain C (to which A itself belongs). B is called the *subgroup generated by C,* and we say briefly: *C generates B.* According to Theorem 4 for example, the set of all transpositions generates the entire symmetric group S_n. Since $(i, j) = (1, i)(1, j)(1, i)$ even the set of all transpositions of the specific form $(1, i)$ generates all of S_n.

Theorem 5. For $n \geq 3$ every element of the alternating group A_n is a product of 3-cycles.

Proof. From the previous remark and Theorem 3 every element of A_n is the product of an even number of transpositions of the form $(1, i)$. Therefore, it suffices to prove the claim for the product of two transpositions of this form. For $i \neq j$ we have $(1, i)(1, j) = (1, i, j)$. The identity is the third power of every 3-cycle, f.i. $I = (1, 2, 3)^3$. □

According to the last proof, A_n is already generated by 3-cycles of the form $(1, i, j)$. Since $(1, i, j) = (1, 2, j)(1, 2, i)(1, 2, j)^2$ even the 3-cycles of the form $(1, 2, i)$ are sufficient.

Example 4. Let $\overline{M}$ be the set of all 1-maneuvers, that is the set of all finite sequences of moves, in which neither two layer moves with the same layer nor two cube moves around the same rotation axis immediately follow each other (cf. section 1.2 and 1. 5). Also the "empty maneuver" which does not contain one single move, and is denoted by $I_{\overline{M}}$, shall belong to $\overline{M}$. We "multiply" two maneuvers by simply writing them next to each other and at the same time combining or cancelling all layer moves with the same layer in immediate succession, as well as all the cube moves around the same axis. Then $\overline{M}$ forms a group, which we call the *large maneuver group.* $I_{\overline{M}}$ is the neutral element of $\overline{M}$, and the maneuver previously denoted by m', the inverse maneuver of $m \in \overline{M}$, is also in the new group-theoretic sense the inverse of m in $\overline{M}$: $m \cdot m' = m' \cdot m = I_{\overline{M}}$. To give a precise, formal definition of the combining and cancelling procedure is easy, but rather tedious; we therefore settle for a few examples:

$$R \cdot R = R^2, \quad R \cdot R^2 = R', \quad LFR^2 \cdot R^2F'B = LB,$$
$$M_R^2 \cdot M_R^2 = I_{\overline{M}}, \quad RC_{URF} \cdot C_{UFR}L = RL.$$

Multiplying maneuvers is a mere game with letters which disregards completely the operations caused by the maneuvers. For example, $M_R^2 M_F^2$ and $M_F^2 M_R^2$ have the same effect, but the two maneuvers are different elements of the large maneuver group: $M_R^2 M_F^2 \neq M_F^2 M_R^2$.

An important subgroup of $\overline{M}$ is the *small maneuver group* denoted by M and consisting of all (outer layer) maneuvers, i.e. all $m \in \overline{M}$ in which neither middle layer moves nor cube moves occur, leaving only outer layer moves.

As a last example we want to look at a group whose full significance for cubology will only become apparent in Section 3.1 and in Chapter 4.

Example 5. Let C be the *rotation group* of the cube, i.e. the group of all "proper distance-preserving transformations" of the space, which bring a given ordinary cube into itself. A distance-preserving transformation of the three-dimensional space is a function of the space into itself, which preserves the distance of any two points. The set of all distance-preserving transformations is a subgroup of the group of all bijections of the space into itself, which we have studied already as Example 3. For any geometric figure the set of all distance-preserving transformations sending the figure as a whole into itself, forms a subgroup of the group of all distance-preserving transformations. It reflects the symmetry properties of the figure: For an asymmetric figure it is "trivial" (i.e. it merely consists of the identity), whereas for a figure with many symmetries it is correspondingly more extensive. A distance-preserving transformation is called *proper* if it, stated somewhat informally, corresponds to a physically possible movement of the space, and this means, it does not require a reflection through a point. Due to the cube moves introduced in section 1.2 and 1.5 we already have a certain understanding of C. For each of the 23 cube moves, $C_R, C_R^2, C_{UF}, C_{ULF}, \ldots$ there is obviously an element $c_R, c_R^2, c_{UF}, c_{ULF}, \ldots$ of C, that means $|C| = 24$, since the identity is to be added. Remarkably, C has exactly the same structure as the symmetric group S_4. This follows from

Theorem 6. *There exists a bijective function π: $C \to S_4$ with $(c_1 \cdot c_2)\pi = c_1\pi \cdot c_2\pi$ for all $c_1, c_2 \in C$.*

The last statement can also be formulated as follows: If $c_1 \cdot c_2 = c_3$ in the group C, then $c_1\pi \cdot c_2\pi = c_3\pi$ in the group S_4. Since π is bijective, it follows conversely that $fg = h$ in S_4 implies $f\pi' \cdot g\pi' = h\pi'$ in C. The groups C and S_4 differ, therefore, only in the names of their elements, not however, in the mutual relations of these elements determined by the operations.

Proof to theorem 6. Every element of C sends the four corner axes of the cube again into corner axes. Therefore, to every element of C a permutation of the 4 corner axes, i.e. an element of S_4, is assigned. It is not difficult to see that this function is bijective and has the required properties. □

Theorem 6 motivates

Definition 4. A function π of a group A into a group B is called a *homomorphism,* if for any $a_1, a_2 \in A$, we have $(a_1 a_2)\pi = a_1\pi \cdot a_2\pi$. If π is also bijective, then π is called an *isomorphism* and the groups A and B *isomorphic* to each other, in symbols: $A \cong B$. A homomorphism or isomorphism of a group into itself is called an *endomorphism* or *automorphism*, respectively.

Theorem 6 now simply becomes $C \cong S_4$.—For every group A and every $a \in A$ the function

$$\pi_a: A \to A, \quad b\pi_a := aba' \quad (b \in A)$$

is an automorphism: It is an endomorphism because $(b_1b_2)\pi_a = ab_1b_2a' = ab_1a' \cdot ab_2a' = b_1\pi_a \cdot b_2\pi_a$, and it is bijective, since to every $b_2 \in A$ there exists exactly one $b_1 \in A$ with $b_2 = b_1\pi_a = ab_1a'$, namely $b_1 = a'b_2a$. The automorphisms of this type are called *inner automorphisms* of the group A. We will play around with them a bit in Section 3.3 (Conjugation). In general, every automorphism describes a kind of structural symmetry of the group, and it transforms every subgroup onto an isomorphic subgroup. Of special interest are subgroups which are always invariant in this process:

Definition 5. A subgroup B of a group A is called a *normal* or *invariant* subgroup of A, if it is invariant under all inner automorphisms, that is if for $a \in A$ and $b \in B$ always $aba' \in B$.

If $A_1, \ldots, A_n$ are any groups, a group structure is defined on the Cartesian product $B := A_1 \times \ldots \times A_n$ in a natural way, namely by componentwise operation: $(a_1, \ldots, a_n) \cdot (\bar{a}_1, \ldots, \bar{a}_n) = (a_1\bar{a}_1, \ldots, a_n\bar{a}_n)$. With this operation B is called the *direct product* of the groups $A_1, \ldots, A_n$. If $A_1 = \ldots = A_n = A$ we again simply write $B = A^n$. If all A_i are abelian, then, obviously, so is B.

For simplicity's sake now let $n = 2$, that is let $B = A_1 \times A_2$ be the direct product of two groups A_1 and A_2. We denote the neutral element of A_1, A_2 and B by e_1, e_2 and e, respectively, and set $A_1^* := \{(a_1, e_2)|a_1 \in A_1\}$, $A_2^* := \{(e_1, a_2)|a_2 \in A_2\}$. It is easily seen, that A_i^* is a subgroup of B isomorphic to A_i $(i = 1, 2)$. More precisely:

(1) A_i^* is a normal subgroup of B $(i = 1, 2)$,
(2) $B = A_1^*A_2^* := \{a_1a_2 | a_1 \in A_1^*, a_2 \in A_2^*\}$,
(3) $A_1^* \cap A_2^* = \{e\}$.

If (1), (2) and (3) hold for arbitrary subgroups A_1^* and A_2^* of a group B, we also call B the (inner) direct product of A_1^* and A_2^* (by identifying A_i^* with A_i), and we write $B = A_1^* \times A_2^*$. If, in this case, A_1^* is a subgroup, but not necessarily normal, we call B the semidirect product of A_1^* with A_2^*.—

2.3 The Small and the Large Model

In Chapter 1 three basic concepts were introduced: "the possible position" (a "color pattern" arrived at from the start position by turning the

nine layers; it remains unaffected by cube moves), "the 1-maneuver" (a sequence of moves, including middle layer and cube moves), and "the possible 1-operation" (a "position change" caused by an 1-maneuver). These three basic concepts correspond to the natural possibilities and the practical requirements of the game. However, put next to each other, they have one small drawback: They don't fit to produce a sensible mathematical model, in which operations are well-defined functions of the set of positions into itself.

It is convenient for such a model to begin by allowing only outer layer moves ("small model"). Thus, the six face cubies always remain in their places serving as a fixed reference system, and a position can easily be described by the location of the corner and edge cubies. For this purpose, we assume that the corners and edges of the cube are numbered from 1 to 8 and 1 to 12, respectively. These numbers are marked both on the mobile cubies and on a fictitious, transparent second skin which loosely surrounds the cube and always stays in place without impeding the rotations of the cube. The numbers on this second skin serve to number the cubicles and coincide in the start position with the numbers on the cubies situated exactly underneath. In any arbitrary position the *place* (not the orientation) of the corner and edge cubies is then defined by an element ρ of the symmetric group S_8 and an element σ of the symmetric group S_{12}: the corner cubie no. i sits in the corner cubicle $i\rho$ ($i = 1, \ldots, 8$), the edge cubie no. j in the edge cubicle $j\sigma$ ($j = 1, \ldots, 12$). In order to describe the *orientation* of the corner and edge cubies, we mark at each of the 8 corners and at each of the 12 edges of the cube one of the three resp. two cube faces meeting at the corresponding corner or edge by an arc or by a line, as f.i. shown in figure 1 above. We do not mark the cubies but the fictitious outer skin of the cube, in other words, we mark one of the three, respectively two outer faces of each cubicle. In addition, we number the three colored faces of each corner cubie clockwise by 0, 1 and 2, and the two colored faces of each edge cubie by 0 and 1, with each 0 lying in the start position beneath the arc or line marking of the corresponding cubicle, cf. figure 1 below. For any arbitrary position the orientation of the corner and edge cubies can now be characterized by an 8-tuple $x = (x_1, \ldots, x_8) \in X := \{0, 1, 2\}^8$ and a 12-tuple $y = (y_1, \ldots, y_{12}) \in Y := \{0, 1\}^{12}$, with x_i or y_j denoting which face of the i-th corner cubie or the j-th edge cubie lies in cubicle i or j beneath the marking. We call the corner cubie no. i *correctly oriented, left twisted* or *right twisted,* if x_i 1=0, 1, or 2, respectively. The edge cubie no. j is called *correctly* or *incorrectly oriented,* if $y_j = 0$ or 1, respectively. (So far such statements only made sense, if the cubie was in its home cubicle.) For our small mathematical model we can now specify the previously somewhat informal concept of the "color pattern":

Definition 1. A position is a quadruple (4-tuple) (ρ, σ, x, y) with $\rho \in S_8$, $\sigma \in S_{12}$, $x = (x_1, \ldots, x_8) \in X := \{0, 1, 2\}^8$ and $y = (y_1, \ldots, y_{12}) \in Y := \{0, 1\}^{12}$. P^* is the set of all positions.

Operations can now be defined as functions of P^* into P^* in an obvious manner. We refrain from further formal definitions and limit ourselves to the following remarks and examples: A *possible operation* is a function of P^* into P^*, "caused" by disassembling and then reassembling the corner and the

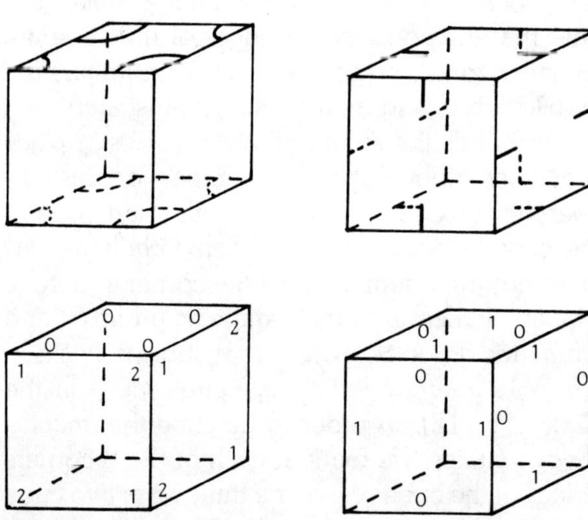

FIGURE 1. Markings of the cubicle faces (above), and numbering of the cubie faces (below)

edge cubies according to a fixed instruction (i.e., according to the same instruction for all $p \in P^*$): The operation $(+urf)(uf, ub)$, for example, assigns a new position to every position by twisting the upper right front corner cubie to the right and swapping the edge cubies from *uf* and *ub* in a certain way. An operation which is not a possible operation is called an *impossible operation*. The set G of the possible operations and the larger set G^* of all operations are subgroups of the symmetric group over P^*. We call *G Rubik's group*. The map

$$\pi: M \rightarrow G$$

which assigns to every maneuver the possible operation it causes, is a homomorphism of the small maneuver group (cf. example 2.2.4) onto Rubik's group: $(m_a m_b)\pi = m_a\pi \cdot m_b\pi$, in words: the product of two maneuvers causes the composition of the possible operations caused by the two maneuvers. Two maneuvers m_a and m_b are called equivalent, denoted by $m_a \sim m_b$, if they cause the same operation: $m_a\pi = m_b\pi$. Due to the function

$$P^* \times M \rightarrow P^*, \ (p, m) \rightarrow pm := p(m\pi),$$

the small maneuver group "acts" on the set of positions; in other words, we have:

(a) $pI_M = p$ for all $p \in P^*$ and
(b) $p(m_a m_b) = (pm_a)m_b$ for all $p \in P^*$ and $m_a, m_b \in M$.

It is immediately seen that

$$p_a \sim p_b, \text{ if and only if there exists an } m \in M \text{ with } p_a = p_b m$$

is an equivalence relation in P^*. The corresponding equivalence classes are called the *orbits* of P^* with respect to M.

Definition 2. A position $p \in P^*$ is called *possible* if it lies in the same orbit as the "start position" $I_P = (1, 1, 0, 0)$, i.e. if there exists a maneuver m with $p = I_P m$. (Here 1 stands for I_{S_8} and $I_{S_{12}}$, 0 for the 8- or the 12-tuple $(0, \ldots, 0)$.) P is the set of all possible positions.

We next turn to the "large model" which considers also middle layer and cube moves. In order to conceive the manipulations of the cube which are now possible as well-defined functions, and to obtain the same mathematical structure as within the "small model", the way the cube is lying on the table must become part of our concept of the position. This way can be determined by defining one of the 6 faces as the upper face and one of the 4 faces, which are then vertical, as the front face. The $6 \cdot 4 = 24$ ways correspond to the elements of the rotation group C of the cube which we have studied as Example 2.2.5 and which is, according to Theorem 2.2.6, isomorphic to S_4. However, to insure as close an analogy as possible to our procedure for corner and edge cubies we describe the way the cube is lying by a permutation of the 6 face cubies, hence by an element $\tau \in S_6$. Of course, only 24 of the $6! = 720$

permutations amount to possible l-positions. The others are "impossible l-positions" which we can only get after taking the cube apart, even the unscrewing of the face cubies is permitted, and then reassembling it. Thus we arrive at

Definition 3. An l-position is a quintuple (5-tuple)$(\rho, \sigma, \tau, x, y)$ with $\rho \in S_8, \sigma \in S_{12}, \tau \in S_6, x \in \{0, 1, 2\}^8$ and $y \in \{0, 1\}^{12}$. $\overline{P}^*$ is the set of all l-positions.

All the other concepts of the "large model" can be explained analogously to those of the "small model": the set $\overline{P}$ of all possible l-positions, the large maneuver group $\overline{M}$ (cf. Example 2.2.4), the group $\overline{G}^*$ of all l-operations, the group $\overline{G}$ of all the possible l-operations, the homomorphism $\overline{\pi}: \overline{M} \to \overline{G}$ and the equivalence of arbitrary l-maneuvers: $m_a \sim m_b$ if and only if $m_a\overline{\pi} = m_b\overline{\pi}$.

The following summary might prove useful:

SMALL MODEL		LARGE MODEL	
M	outer layer maneuvers	$\overline{M}$	all maneuvers
P, P^*	possible resp. all positions	$\overline{P}, \overline{P}^*$	possible resp. all 1-positions
G, G^*	possible resp. all operations	$\overline{G}, \overline{G}^*$	possible resp. all 1-operations
$\pi: M \to G$	homomorphism	$\overline{\pi}: \overline{M} \to \overline{G}$	homomorphism

2.4 Characterization of Possible Positions and Operations

The following basic theorem allows us to determine whether any arbitrary position can or cannot be transformed back into the start position merely by turning layers.

Theorem 1 ("first law of cubology"). A position (ρ, σ, x, y) is possible, if and only if the following three conditions are fulfilled.

(a) $\operatorname{sgn} \rho = \operatorname{sgn} \sigma$,

(b) $x_1 + x_2 + \ldots + x_8 = 0$ modulo 3,

(c) $y_1 + y_2 + \ldots + y_{12} = 0$ modulo 2.

Proof.

(1) We first demonstrate that the three conditions are *necessary*, i.e. that they hold for every possible position. This is not particularly difficult. On one

hand, (a), (b) and (c) are fulfilled by the start position I_P with $\rho = I_{S_8}$, $\sigma = I_{S_{12}}$, hence sgn ρ = sgn $\sigma = 1$, and with $x_1 = x_2 = \ldots = x_8 = y_1 = y_2 = \ldots = y_{12} = 0$. On the other hand, (a), (b) and (c) are preserved in every one of the six 90°-moves *U, D, R, L, F* and *B*: (a) remains valid, since every one of these moves simultaneously causes a corner-4-cycle and an edge-4-cycle, hence an odd permutation of the corner cubies and an odd permutation of the edge cubies. (b) remains valid, because the components of *x* with *U* or *D* do not change at all, while with *R, L, F* and *B* simultaneously two components are increased by 1 modulo 3, and two components are reduced by 1 modulo 3 (figure 2.3.1, left). (c) remains valid, because with each of the six moves exactly four components of *y* are changed by 1 (figure 2.3.1, right).

(2) In order to prove that the three conditions are *sufficient,* we have to show, that every position $p = (\rho, \sigma, x, y)$ fulfilling the three conditions can be transformed back into the start position by means of a suitable maneuver *m*, in short, that there exists for each such *p* an $m \in M$ with $pm = I_P$. As a matter of fact, this was already done by describing the "simple strategy". But we want to give here an independent and slightly more abstract proof.

(α) Without loss of generality we can assume sgn ρ = sgn $\sigma = 1$. Namely, if sgn ρ = sgn $\sigma = -1$, we have for $p_a := (\rho_a, \sigma_a, x_a, y_a) := pU$ the equation sgn ρ_a = sgn $\sigma_a = 1$, and from $m_a \in M$ with $p_a m_a = I_P$ it follows $p(Um_a) = (pU)m_a = I_P$.

(β) Let us now consider any arbitrary maneuver for a corner-3-cycle, for example $m_{100} = RB'RF^2R'BRF^2R^2$ (9), which causes the 3-cycle (X_1, X_2, X_3) shown in figure 1.5.2 above on the left for the corners X_1 = *ufl*, X_2 = *urf*, X_3 = *ubr*, without changing the other corners, which we denote for the time being by $X_4, \ldots, X_8$. For every $i \in \{4, \ldots, 8\}$ there exists a maneuver $k_i \in M$ of at most 2 moves which transports the corner cubie from X_i to X_3 without interfering with the cubies in X_1 and X_2. The maneuver $k_i m_{100} k_i'$ now causes exactly the 3-cycle (X_1, X_2, X_i). (The principle of "conjugation" used here will be discussed more extensively in Section 3.3.) Since the six 3-cycles (X_1, X_2, X_3), $(X_1, X_2, X_4), \ldots, (X_1, X_2, X_8)$ generate the group of all even permutations of $\{X_1, \ldots, X_8\}$ (cf. remark after Theorem 2.2.5) there exists a maneuver $m_C \in M$, which transports all eight corner cubies into their home cubicles.

(γ) Analogously, proceeding from an arbitrary maneuver for an arbitrary edge-3-cycle, we can find a maneuver m_E, which brings all the twelve edge cubies into their proper places.

(δ) In position $pm_C m_E$ ($= pm_E m_C$) all the corner and edge cubies are already in their correct cubicles. According to part (1) of the proof the equations (b) and (c) are also valid for this position, i.e. an even number of edge cubies needs reorientation and the number of corner cubies to be twisted to the right is equal to the number of corner cubies to be

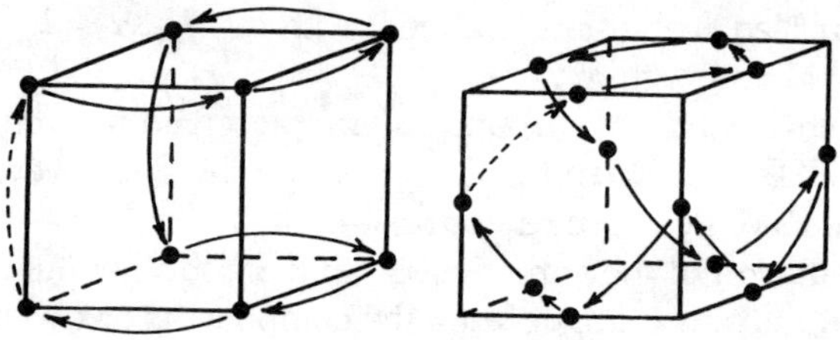

FIGURE 1. Paths around the cube which connect neighboring corner cubies resp. neighboring edge cubies and reach each corner resp. edge cubie exactly once.

twisted to the left modulo 3. We, therefore, arrive at the correct orientation of all the corner and edge cubies by

$$LFR'F'L'U^2RURU'R^2U^2R \ (13) \rightarrow (+uf)(+ur)$$

and

$$R(U^2RF'D^2FR')^2R' \ (13) \rightarrow (+urf)(-ufl)$$

(as well as trivial "variations" of these maneuvers, cf. Section 3.1). Here we can, for instance, follow "step-by-step" a path around the cube which connects neighboring corner respectively edge cubies and reaches each corner respectively edge cubie exactly once, cf. figure 1. □

An immediate consequence of our first law of cubology is

Theorem 2. *The number of possible positions is*

$$|P| = |G| = \frac{1}{12}|P^*| = \frac{1}{12} \cdot 8! \cdot 12! \cdot 3^8 \cdot 2^{12}$$
$$= 43\ 252\ 003\ 274\ 489\ 856\ 000 = 2^{27} \cdot 3^{14} \cdot 5^3 \cdot 7^2 \cdot 11.$$

Proof. For the 8 corner cubies there are 8!, for the 12 edge cubies there are 12! arrangements (permutations). Since every corner cubie can at every corner be oriented in 3, and every edge cubie on every edge in 2 different ways, we have

$$|P^*| = 8! \cdot 12! \cdot 3^8 \cdot 2^{12}.$$

For *possible* positions this number is reduced to a half because of equation (a) (there are as many even as odd permutations), to a third because of equation (b) (the orientation of 7 corner cubies can be arbitrarily chosen and determines the orientation of the 8th corner cubie), and again to a half because of equation (c) (the orientation of 11 edge cubies can be arbitrarily chosen and determines the orientation of the 12th edge cubie), all in all it is therefore reduced to one twelfth. □

Curiously and quite untypically, manufacturer and merchants promoting Rubik's cube in innumerable posters and advertisements are satisified with the modest slogan "the magic cube with its 43 trillion different faces". In reality, the cube has more than 43 *quintillion* faces, which is, after all, more than one million times more!

If we put 43.252 quintillion cubes of 5.6 cm width–each in another position–side by side, they bridge a distance of $2.422 \cdot 10^{15}$ (2.422 quadrillion) km, this is about 256 light-years. (To compare: The principal stars α, β, γ and δ of the Great Bear are respectively 96, 76, 80 and 76 light-years away from our solar system. The North star, the principal star α of the Little Bear, is 470 light-years away.) Tightly packed the cubes would cover the whole surface of the earth (land and water, $51 \cdot 10^{17}$ cm^2) to a height of 15 m.

Despite these perhaps impressive comparisons, the main reason for the difficulty of our game is *not* the size of the number $|P|$. It is certainly not considered difficult, to arrange 21 cards of a customers list in alphabetical order. Yet, for 21 cards there are $21! = 51.09 \ldots \cdot 10^{18}$ (over 51 quintillion) arrangements, and if all the names are different, only one of them corresponds to the desired alphabetical order. The indisputable difficulty of Rubik's cube is rather due to the intricate way in which 8 to 9 cubies at a time are linked, forcing us to head for the solution by way of sophisticated detours. Notice that immediately before the last move 8 cubies, and immediately before the second to last

move 12 to 16 cubies (12 only in the case of two 180°-middle-layer-moves) must necessarily be wrong, therefore more than half of all the 26 visible cubies.

Between the operations $g \in G^*$ and the positions $p \in P^*$ there exists the natural bijection $G^* \ni g \leftrightarrow p = I_P g \in P^*$, in which p is a possible position if and only if g is a possible operation. Theorem 1, which characterizes the possible positions, can therefore be directly "translated" into a theorem describing the possible operations. A brief preliminary remark: The rth power of a corner-r-cycle with sign "+" resp. "−" obviously causes a right- resp. left-twist for all r corner cubies of the cycle. The rth power of an edge-r-cycle with sign "+" reorients all r edge cubies of the cycle. Consequently, the sign of a cycle is independent of its representation, and we can distinguish between *orientation-preserving corner cycles, right-twisting corner cycles, left-twisting corner cycles, orientation-preserving edge cycles* and *reorienting edge cycles.* Thus, the previously announced "translation" of Theorem 1 results in

Theorem 3 ("second law of cubology"). An operation is possible, if and only if the following three conditions are fulfilled:

(a) The total number of cycles of even length (corner and edge cycles) is even.

(b) The number of right-twisting corner cycles is equal to the number of left-twisting corner cycles modulo 3.

(c) The number of reorienting edge cycles is even.

Proof. Equation (a) of Theorem 1 says that the permutation caused in the 20-element set of all corner and edge cubies is even. This is equivalent to condition (a) of Theorem 3, since an individual cycle is an even permutation if and only if its length is odd.

For the equivalence of the conditions (b) the statement suffices that a right-twisting, left-twisting or orientation-preserving corner cycle changes the sum of the orientation coordinates x_i (summation over all the cubies in the cycle) by -1, $+1$ or 0 modulo 3, respectively. Analogously, we have for the conditions (c): An edge cycle is reorienting, if and only if the sum of the orientation coordinates y_j of the cubies in the cycle changes by an odd number. □

A simple corollary of Theorem 3 is

Theorem 4. Every position derived from a possible position by swapping two corner cubies (arbitrary orientation), swapping two edge cubies (arbitrary orientation), twisting an individual corner cubie or reorienting an individual edge cubie, is impossible.

S. W. Golomb (University of Southern California) has pointed out a remarkable analogy between the behavior of the corner cubies which can only be twisted in pairs into different directions and in threes in the same direction, and the hypothetical quarks of elementary particle physics. Quarks and their antiparticles, the antiquarks, seem unable to exist in isolation, but appear only as quark-antiquark-pair in the form of a meson, as quark-triplet in the form of a baryon and as antiquark-triplet in the form of an antibaryon. Will we perhaps–one day–be able to give a group theoretical explanation for this, too? Or–a frivolous speculation–will the day come, when new elementary particles are looked for and actually found on the basis of the properties of our edge cubies?—

2.5 The Structure of Rubik's Group

There is the natural bijection $G \ni g \rightarrow I_P g \in P$ between Rubik's group G and the set P of all possible positions. We can therefore identify every group element g with the position $I_P g = (\rho, \sigma, x, y) \in P$. Here, the corner cubie no. i is sitting in cubicle $i\rho$, and beneath the arc marking of this cubicle we have the orientation coordinate $x_i = ix$ ($i = 1, \ldots, 8$). If we then apply a second group element $g^* = (\rho^*, \sigma^*, x^*, y^*) \in G$, the corner cubie no. i moves into cubicle $i\rho\rho^*$, and beneath the arc marking of this cubicle we now have the number $ix + i\rho x^*$ modulo 3. Analogously, edge cubie no. j moves into the edge cubicle $j\sigma\sigma^*$ with the number $jy + j\sigma y^*$ modulo 2 beneath the line marking of this cubicle. By taking the sets $X = \{0, 1, 2\}^8$ and $Y = \{0, 1\}^{12}$ as the direct product of 8 cyclic groups C_3 respectively 12 cyclic groups C_2, i.e. by defining the addition in X respectively Y componentwise modulo 3 respectively 2, we get for the product of two elements of Rubik's group G the basic equation

$$(\rho, \sigma, x, y)(\rho^*, \sigma^*, x^*, y^*) = (\rho\rho^*, \sigma\sigma^*, x + \rho x^*, y + \sigma y^*).$$

With this, many structural properties of G can be formally deduced. Let

$$G_1 := \{g = (\rho, \sigma, x, y) \in G \mid x = 0, y = 0\},$$
$$G_2 := \{g = (\rho, \sigma, x, y) \in G \mid \rho = 1, \sigma = 1\}.$$

For brevity and clarity we again write 0 for the 8- resp. 12-tuple $(0, \ldots, 0)$ and 1 for I_{S_8} respectively $I_{S_{12}}$. Hence, G_1 is the set of the possible operations preserving the orientation of all the cubies (with regard to the chosen marking), G_2 is the set of the possible operations leaving all the cubies in their cubicles.

Theorem 1.

(a) G_1 is a subgroup, G_2 a normal subgroup of G.

(b) $G_1 \cong \{(\rho, \sigma) \in S_8 \times S_{12} \mid sgn\ \rho = sgn\ \sigma\}$, $G_2 \cong C_3^7 \times C_2^{11}$.

(c) G is the semidirect product of G_1 with G_2.

Proof.

(a) The subgroup property of G_1 and G_2 is trivial, and the invariance of G_2 quickly checked: For $g = (\rho, \sigma, x, y) \in G$ we have $g' = (\rho', \sigma', \rho'(-x), \sigma'(-y))$, and for any arbitrary $g^* = (1, 1, x^*, y^*) \in G_2$ it follows $gg^*g' = (1, 1, \rho x^*, \sigma y^*) \in G_2$.

(b) Trivial. ($G_2 \ni (1, 1, (x_1, \ldots, x_8), (y_1, \ldots, y_{12})) \to ((x_1, \ldots, x_7), (y_1, \ldots, y_{11})) \in C_3^7 \times C_2^{11}$ is an isomorphism.)

(c) Due to (a) we only have to check that $G_1 \cap G_2 = \{I_G\}$ and $G_1G_2 = G$. The last equation holds because of $(\rho, \sigma, x, y) = (\rho, \sigma, 0, 0) \cdot (1, 1, \rho'x, \sigma'y)$. □

The following theorem demonstrates the extreme non-commutativity of G:

Theorem 2. The center $Z(G)$ of Rubik's group merely consists of I_G and the superflip which is caused by our favorite maneuver m_{490} and reorients all twelve edge cubies.

Proof. Let $g = (\rho, \sigma, x, y) \in Z(G)$. Since the center of the symmetric group S_n for $n \geq 3$ is trivial (Theorem 2.2.2) and since every $\rho^* \in S_8$ appears as a first coordinate of an element of G, it immediately follows from our fundamental product equation $\rho = 1$ and analogously $\sigma = 1$, i.e. $g \in G_2$. Thus, $gg^* = g^*g$ for all $g^* \in G$ simply becomes $x + x^* = x^* + \rho^*x$, i.e. $x = \rho^*x$ for all $\rho^* \in S_8$, and $y + y^* = y^* + \sigma^*y$, i.e. $y = \sigma^*y$ for all $\sigma^* \in S_{12}$, and this means, x and y are constant. The first law of cubology excludes the cases $x = 1$ and $x = 2$. This leaves only the two elements $I_G = (1, 1, 0, 0)$ and $(1, 1, 0, 1)$ (superflip), which obviously do belong to $Z(G)$. □

The interchangeability of the superflip with other operations can be demonstrated, for instance, like this: We take a cube in the start position, begin by applying the superflip with m_{490}, follow this up with one or two arbitrary moves executed behind our back, apply the superflip again (preferably also behind our back) and can then undo the intermediate moves, made 24 layer moves ago.

Apart from the center, the so-called *commutator subgroup* $K(A)$ of a group A is a measure of the commutativity of A. It consists of all finite products of *commutators*, i.e. of elements of the form $[a, b] := aba'b'$.

It is easily seen that $K(A)$ is always a subgroup, and even a normal subgroup, which, in the case of an abelian group, contains only the neutral element. But in our case $K(G)$ is "half the group":

Theorem 3. $K(G) = \{(\rho, \sigma, x, y) \in G \mid sgn\ \rho = sgn\ \sigma = 1\}$.

Proof. If $g = [g_1, g_2] = g_1g_2g_1'g_2' = (\rho, \sigma, x, y)$ is a commutator with $g_1 = (\rho_1, \ldots)$ and $g_2 = (\rho_2, \ldots)$, then $\rho = \rho_1\rho_2\rho_1'\rho_2'$ holds, and hence $sgn\ \rho = 1$ (Theorem 2.2.3). The condition $sgn\ \rho = sgn\ \sigma = 1$, therefore, holds for all commutators, hence for all products of commutators (again Theorem 2.2.3), i.e. for all $g \in K(G)$. Conversely, every $g = (\rho, \sigma, x, y) \in G$ with $sgn\ \rho = sgn\ \sigma = 1$ is the product of commutators, as follows from part (2), $(\beta) - (\delta)$, of the proof of the first law of cubology (Theorem 2.4.1): For the elementary operations used in this proof (corner-3-cycle, edge-3-cycle, twisting of two corner cubies, reorienting of two edge cubies), there exist maneuvers possessing the form of a commutator of the small maneuver group M, for example, m_{101}, m_{551a}, m_{5a} and

$$FUD'L^2U^2D^2R \cdot U \cdot R'D^2U^2L^2DU'F' \cdot U' \rightarrow (+uf)(+ur).$$

Conjugates (cf. Section 3.3) of commutators are commutators again since

$$m \cdot m_a m_b m_a' m_b' \cdot m' = (mm_a m')(mm_b m')(mm_a m')'(mm_b m')'.$$

As image element under the homomorphism π the operation caused by a product of commutators in M is a product of commutators in G. □

Due to $m_{990} = RL'F^2B^2RL' \cdot U \cdot LR'B^2F^2LR'$ (13) $\sim D$ (D. J. Benson and others), G is already generated by the five operations caused by R, L, F, B and U. Yet we cannot do without *two* of the six moves R,

L, F, B, U and *D:* If the "forbidden layers" are mutually perpendicular, one edge cubie is immovable. If they are opposite to each other, as f.i. *U* and *D*, it is impossible to reorient edge cubies. This is already the case, if only 180°-moves are allowed for two opposite layers, like *U* and *D,* while the other four outer layers can be turned arbitrarily: It is obvious that f.i. the upper colored tile of the edge cubie *uf* can never reach the front or the back side of the cube. In other words (and this shows the practical value of subgroup investigations demonstrated in the next section): Every maneuver which reorients edge cubies contains at least three 90°-outer-layer-moves, namely for each of the three pairs of opposite outer layers one 90°-move with one of the two layers.

The symmetric group S_n is generated already by two elements, as f.i. the two cycles (1, 2) and (2, . . . , n). (For $2 \leq i \leq n$ we have $(1, i) = (2, \ldots, n)^{n+1-i}(1, 2)(2, \ldots, n)^{i-2}$, and the transpositions of the form $(1, i)$ generate S_n, as was shown in the remarks following Theorem 2.2.4.) An analogous two-element set of generators for G including brief maneuvers has been given by F. Barnes:

m_{991} = *UBLUL'U'B'* (7) → (−*ufl, ubr*)(+*ulb*)(+*ub, ur*)(+*ul*),
m_{992} = *R²FLD'R'* (5) → (*urf, bdr, lfd, ldb, lbu, luf, frd*)
(*uf, rd, rf, ru, fd, db, br, ld, lb, lu, lf*).

There exist other results about the structure of G and G^*, but they would go beyond the scope of this book's mathematical content which should not overstrain the patience of non-mathematicians. We only mention normal series, representations as wreath products (f.i. $G^* = (S_8 \wr C_3) \times (S_{12} \wr C_2)$), generating G by elements of the lowest possible order and so on.

2.6 Special Subgroups

The subgroup structure of Rubik's group G is extremely varied.

The easiest way, for the present, is to find all the cyclic subgroups of G. A group is called *cyclic,* if it is generated by one single element. Since every finite cyclic group of the order n is isomorphic to C_n (Example 2.2.2) and every infinite cyclic group is isomorphic to the additive group of integers (Example 2.2.1) we already know the structure of all cyclic groups. By the *order* of an element a of a group A we mean the order of the cyclic subgroup generated by a. In the case of a finite group this is the smallest natural number n with $a^n = e$ (neutral element). For Rubik's group the order of all the elements can be immediately read off the cyclic decomposition: It is the least common multiple of the cycle lengths multiplied by 3 (twisting corner cycles), or by 2 (reorienting edge cycles), or by 1 (orientation-preserving cycles). There exist precisely 73 different orders and the maximum order is $2 \cdot 2 \cdot 3 \cdot 3 \cdot 5 \cdot 7 = 1260$. The following short maneuver for an operation of this order has been found by J. B. Butler:

$$RU^2D'BD' \ (5) \rightarrow (-ufl,\ lbu,\ rfu)(+ubr,\ fdl,\ dfr,\ rbd,\ ldb)$$
$$(+uf,\ lb,\ dr,\ fr,\ ul,\ ur,\ bu)(+dl,\ rb)(df,\ db).$$

By the way, 1260 is also the maximum order in the group $\overline{G}$, and here a maneuver with one single layer move is already sufficient:

$$R|C_U \ (1) \rightarrow (+ufl,\ ulb,\ ubr,\ rdf)(+urf)(+dlf,\ dbl,\ drb)$$
$$(+uf,\ ul,\ ub,\ ur,\ rf)(+df,\ dl,\ db,\ dr,\ lf,\ bl,\ rb)(f,\ l,\ b,\ r).$$

In general, we are particularly interested in subgroups defined by either the requirement not to move certain cubies, or to move them only in a restricted way, or by a restriction to certain moves or maneuvers. It follows from the second law of cubology (Theorem 2.4.3) that the structure of the subgroup of all possible operations which leave a certain subset C of the set of all corner cubies and a certain subset E of the

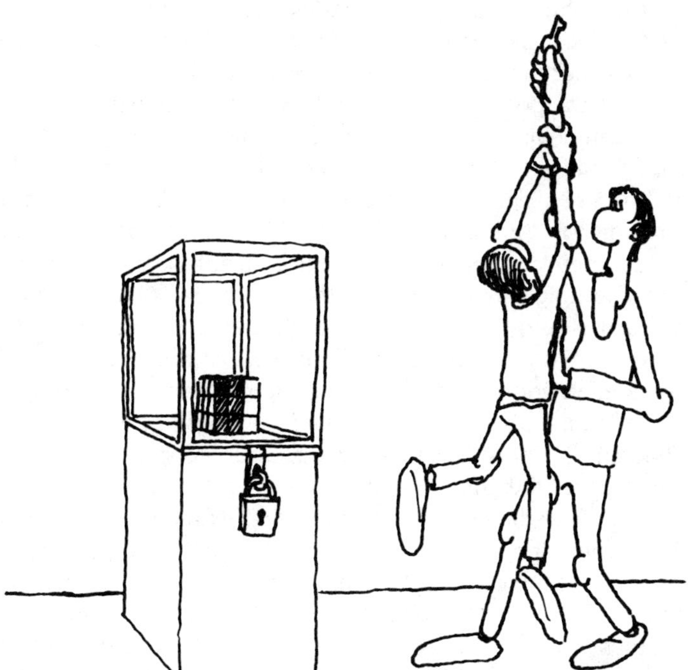

set of all edge cubies untouched (elementwise fixed), does not depend on the location but only on the number of the corner and edge cubies remaining untouched. With $c := 8 - |C|$ and $e := 12 - |E|$, such a subgroup has the order $(c!e!3^c2^e)/12$. As an example of a subgroup defined by a restriction to certain moves, we look at the "square group" $\langle\langle R^2, L^2, F^2, B^2, U^2, D^2\rangle\rangle$ generated by the operations of the six square moves $R^2, L^2, F^2, B^2, U^2, D^2$. (The inner brackets are supposed to indicate the transition from the maneuvers to the operations, i.e. the homomorphism π, while the outer brackets indicate the transition to the generated subgroup). As already frequently done, we identify every operation g with the position $I_P g$, which is obtained by applying g to the start position I_P. We call a cubie red or blue etc., if one of its color tiles is red or blue etc. Colors sitting opposite each other in the start position are called "counter colors". With this terminology we have

Theorem 1. The square group $Q := \langle\langle R^2, L^2, F^2, B^2, U^2, D^2\rangle\rangle$ has the order $2^{13} \cdot 3^4 = 663\,552$. A position lies in Q, if and only if (a) on each of the six cube faces there appears only the color of the face cubie and its counter color and (b) the centers of the four corner cubies (considered as small cubes) of any fixed color lie in one plane.

Proof. The necessity of the conditions (a) and (b) is trivial: They hold for the start position and are preserved with each of the six square moves. To ascertain whether the conditions are sufficient, we have to examine more precisely which of the positions can be reached. To begin with, in every cubicle every cubie can obviously have only one orientation. Moreover, the set of the eight corner cubies is divided into two disjoint "tetrahedrons" $\{urf, ulb, drb, dlf\}$ and $\{ufl, ubr, dfr, dbl\}$, and the set of the twelve edge cubies is divided into three disjoint "rings" $\{uf, ub, db, df\}$, $\{ul, ur, dr, dl\}$, $\{fl, fr, br, bl\}$, in other words, the elements of each of these five 4-element sets can only be permuted among themselves. After that, there are still $4!^2$ placements of the corner cubies and $4!^3$ placements of the edge cubies conceivable. However, since every square move simultaneously causes a corner-double-transposition and an edge-double-transposition, the permutation of the corner cubies as well as the permutation of the edge cubies has to be even. (According to Theorem 2.5.3 we are, therefore, in a subgroup of the commutator group.) This reduces the number of placements of the edge cubies to $(4!^3)/2$. For the corner cubies the last restriction together with condition (b) means that after determining the places of the four cubies of one tetrahedron, there remain only four possibilities for those of the other, hence $4! \cdot 4$ possibilities for the 8 corner cubies in all. Finally, all the arrangements of corner and edge cubies enumerated here are also possible independent of each other; thus the number of positions obtainable by means of square moves amounts to

1 Disassembled Magic Cube

2 The Magic Domino

3 Left: “Six Dots”
Right: “Chessboard Cube of Order 2”

4 Left: “Chessboard Cube of Order 3”
Right: “Chessboard Cube of Order 6”

5 Left: "Christmas Parcel of Order 2"
Right: "Christmas Parcel of Order 3"

6 Left: "Four Plus"
Middle: "Six Plus-Dot-Faces"
Right: "Four Plus and Two Chessboard Faces"

7 Left: "Twisted Peaks"
Right: "Twisted Rings"

8 Left: "Exchanged Peaks"
Right: "Exchanged Rings"

9 Left: "Twisted Chicken Feet"
Right: "Twisted Duck Feet"

10 Left: "Exchanged Chicken Feet"
Right: "Exchanged Duck Feet"

11 Left: "Cubes within the Cube"
Right: "Anaconda"

12 Left: "Four Diagonals"
Middle: "Four Diagonals and Two Plus"
Right: "Six Diagonals" (impossible)

13 Left: "Anaconda"
Right: "Python"

14 Left: "Edge Hexagon of Order 3"
Right: "Edge Hexagon of Order 2"

15 Left: "Six U"
Right: "Six C"

16 Left: "Six T"
Right: "Six Double L"

17 Left: "Gay Rings"
Right: "Anaconda multicolor 1"

18 Left: "Six Tricolours"
Right: "Rescue Cube"

19 Left: "Superflip"
Middle: "Supertwist"
Right: "Superfliptwist"

20 Left: 2x2x2-Cube
Middle: 5x5x5-Cube
Right: Mèffert's Pyraminx

$4! \cdot 4 \cdot (4!^3)/2 = 2 \cdot (4!)^4 = 2^{13} \cdot 3^4$. This is shown by the following maneuvers:

$$q_1 = F^2U^2R^2 \ (3) \to (ufl, ubr) \cdot (urf, dlf, ulb, drb)(uf, df, ub)(ul, dr, ur)(fl, br, fr),$$
$$q_2 = q_1U^2 = F^2U^2R^2U^2 \ (4) \to (urf, dlf)(ulb, drb) \cdot (uf, df)(ul, dr)(fl, br, fr),$$
$$q_3 = m_{620} = (R^2U^2)^3 \ (6) \to (rf, rb)(uf, ub),$$
$$q_4 = (D^2B^2D^2R^2)^2 \ (8) \sim m_{500} \to (fr, bl, br).$$

By means of q_1 (or trivial variations of q_1, cf. Section 3.1) one of the two tetrahedrons can be arbitrarily arranged. The four arrangements which are then possible for the other can be obtained by means of q_2. After, at most, one application of q_3, on each of the three rings an even permutation is required, and this is brought about by q_4. □

The following maneuver demonstrates that condition (b) in Theorem 1 is essential:

$$U^2BU^2B' \cdot R^2B'R^2B \cdot R^2FR^2F' \ (12).$$

It transforms the start position into a position which fulfills condition (a) but does not belong to Q.

Due to

$$R^2F^2B^2L^2 \cdot U^2 \cdot L^2B^2F^2R^2 \ (9) \sim D^2,$$

to generate Q one of the six square moves is superfluous, and we have, for example,

$$\langle\langle R^2, L^2, F^2, B^2, U^2, D^2\rangle\rangle = \langle\langle R^2, L^2, F^2, B^2, U^2\rangle\rangle.$$

For a number of further subgroups of this kind we merely indicate the order where we simply write $|\ldots|$ instead of $|\langle\langle\ldots\rangle\rangle|$:

$\lvert R^2\rvert$	$= 2$	=	2
$\lvert R\rvert$	$= 2^2$	=	4
$\lvert R^2, L^2\rvert$	$= 2^2$	=	4
$\lvert R^2, L\rvert$	$= 2^3$	=	8
$\lvert R, L\rvert$	$= 2^4$	=	16
$\lvert R^2, U^2\rvert$	$= 2^2 3$	=	12
$\lvert R^2, U\rvert$	$= 2^6 3^2 5^2$	=	14 400
$\lvert R, U\rvert$	$= 2^6 3^8 5^2 7$	=	73 483 200
$\lvert R^2, L^2, U^2\rvert$	$= 2^5 3$	=	96
$\lvert R, L, U\rvert$	$= 2^{14} 3^{13} 5^3 7^2$	=	159 993 501 696 000
$\lvert R^2, F^2, U^2\rvert$	$= 2^5 3^4$	=	2 592
$\lvert R, F, U\rvert$	$= 2^{18} 3^{12} 5^2 7^2$	=	170 659 735 142 400
$\lvert R^2, L^2, U^2, D^2\rvert$	$= 2^6 3$	=	192

$$\begin{array}{lllr}
|R^2, L^2, U^2, F^2| & = 2^{11}3^4 & = & 165\,888 \\
|R^2, L^2, F^2, B^2, U^2, D^2| & = 2^{13}3^4 & = & 663\,552 \\
|R^2, L^2, F^2, B^2, U, D| & = 2^{16}3^5 5^2 7^2 & = & 19\,508\,428\,800 \\
|R^2, L^2, F, B, U, D| & = 2^{16}3^{14}5^3 7^2 11 & = & 21\,119\,142\,223\,872\,000 \\
|R, L, F, B, U, D| & = 2^{27}3^{14}5^3 7^2 11 & = & 43\,252\,003\,274\,489\,856\,000
\end{array}$$

The maneuver

$$(RL'F^2R^2L^2FR^2L^2B'R^2L^2F'RL' \cdot C_U)^2C_U^2 \ (28) \sim U^2 \quad \text{(P. Klerings)}$$

shows that

$$\langle\langle R^2, L^2, F, B, U, D\rangle\rangle = \langle\langle F, B, U, D\rangle\rangle.$$

It is particularly fascinating to "realize" a given finite group A "on the cube", i.e. to find a subgroup of G isomorphic to A and, if possible, brief maneuvers for its elements. This can be done for all groups of the order <13; the smallest abelian group which cannot be realized on the cube is the cyclic group C_{13}, and the smallest non-abelian group which cannot be realized on the cube is the so-called *dihedral group* of order 26 (rotation group of a regular polygon of 13 vertices in three-dimensional space, cf. Example 2.2.5). Here too, we limit ourselves to one example: The eight possible operations

To be or not to be . . .

$$
\begin{aligned}
1 &:= I_G, \\
-1 &:= (+uf)(+ul)(+ub)(+ur) && \text{(cf. } m_{435}\text{)}, \\
i &:= (+ur,\ uf)(+ul,\ ub) && \text{(cf. } m_{706}\text{)}, \\
j &:= (+ul,\ uf)(+ub,\ ur) && \text{(cf. } m_{707}\text{)}, \\
k &:= (+uf,\ ub)(+ul,\ ur) && \text{(cf. } m_{710}\text{)}, \\
-i &:= i',\ -j := j',\ -k := k'
\end{aligned}
$$

form a subgroup of G, in which the computing rules

$$
\begin{aligned}
&i^2 = j^2 = k^2 = -1, \\
&ij = (-1)ji = k,\ jk = (-1)kj = i,\ ki = (-1)ik = j
\end{aligned}
$$

hold. It is called the *quaternion group* and plays a somewhat exotic role in group theory. It is, for example, the smallest non-abelian group whose subgroups are all normal.

3 CONSTRUCTION OF MANEUVERS

What is a "beautiful maneuver"?—Since every maneuver has to result in a given operation, it is judged according to its performance, the primary requirement being effectiveness: The maneuver should produce its operation in the fewest possible layer moves. Middle layer moves, usually performed in two steps, are less "beautiful" than outer layer moves, and 180°-moves are less "beautiful" than 90°-moves. Since every maneuver must be stored in our brain and flow from our finger tips, a catchy and handy structure is desirable; in most cases this means moving as few layers as possible, and these layers should be easily accessible for right-handed people—without loss of generality.

This chapter contains four useful methods for the construction of "beautiful maneuvers".

3.1 Variation

We begin with a simple example. If one is familiar with maneuver.

$$m_{101} = R \cdot BL'B' \cdot R' \cdot BLB' \quad (8) \rightarrow (\mathit{ufl},\ \mathit{urf},\ \mathit{bru})$$

(figure 1.5.2, top center), neither the inverse maneuver

$$m_{101} = BL'B' \cdot R \cdot BLB' \cdot R' \quad (8) \rightarrow (\mathit{ufl},\ \mathit{bru},\ \mathit{urf})$$

nor a maneuver like

$$B \cdot LF'L' \cdot B' \cdot LFL' \ (8) \rightarrow (\mathit{urf},\ \mathit{ubr},\ \mathit{lbu})$$

appear particularly ingenious. The last maneuver results from m_{101} by everywhere replacing R by B, B by L, and L by F, thus it is obviously equivalent to $C_U m_{101} C_U'$. A somewhat less trivial case is the "reflection" of m_{101} through the plane containing the cube edges *fr* and *bl*: Here R changes to F', B to L', etc., and we get

$$F' \cdot L'BL \cdot F \cdot L'B'L \ (8) \rightarrow (\mathit{ufl},\ \mathit{bru},\ \mathit{rfu}).$$

The subsequent inversion results in

$$L'BL \cdot F' \cdot L'B'L \cdot F \ (8) \to (ufl,\ rfu,\ bru).$$

Thus, we have a new maneuver for the operation shown in figure 1.5.2 center, which differs in only one cycle element from the one we started from caused by m_{101}.

In order to understand the previously indicated possibilities in full generality, we need the concept of the *symmetry group* S of the cube. It results from an extension of the rotation group C (cf. Section 2.2) by the so-called improper symmetries of the cube. These are the distance-preserving transformations of the three-dimensional space which bring the cube as a whole back onto itself, without, however, allowing the axis rotations or the identity already covered by C. Without going into details we observe: S contains 48 elements, namely the 24 elements of C and 24 elements of the form cs_M, c being an element of C and s_M being the reflection through the center of the cube. S is the direct product of the two normal subgroups C and $\{I_S, s_M\}$.

The various types of transformations belonging to S are illustrated in figure 1 by showing their effects on the 8 corners of the cube: The reflection s_M maps every corner onto the corner diagonally opposite (figure 1 above right). If c is a 180°-turn around a face axis or an edge axis then cs_M is the reflection through the perpendicular bisecting plane of the section of the rotation axis lying inside the cube (figure 1, third and fourth row). If c is a 90°-turn around a face axis, then cs_M is a transformation of order 4 (figure 1, second row). Finally, if c is a turn around a corner axis, then cs_M is a transformation of order 6 (figure 1, last row).

Due to the perfect symmetry properties of Rubik's cube, every $s \in S$ is assigned a bijection of the set of moves in a natural manner. The various types of moves (90°-outer-layer-moves, 180°-outer-layer-moves, 90°-middle-layer-moves, 180°-middle-layer-moves, 90°-face-axis-turns, 180°-face-axis-turns, edge-axis-turns, corner-axis-turns) are always permuted among themselves. Instead of a formal definition we give a few examples, writing for the image of a move z under s briefly zs:

$$\begin{aligned} &Rc_U = F, \quad Rc_{UF} = L, \quad Rs_M = L', \\ &M_Rc_U = M_F, \quad M_Rc_{UF} = M_R', \quad M_Rs_M = M_R, \\ &C_Uc_U = C_U, \quad C_{UL}c_{UF} = C_{FR}, \quad C_{URF}s_M = C_{URF}. \end{aligned}$$

Applied literally "move by move", as f.i.

$$\begin{aligned} (BM_RB^2M_R'B)s_M &= (Bs_M)(M_Rs_M)(B^2s_M)(M_R's_M(Bs_M) \\ &= F'M_RF^2M_R'F', \end{aligned}$$

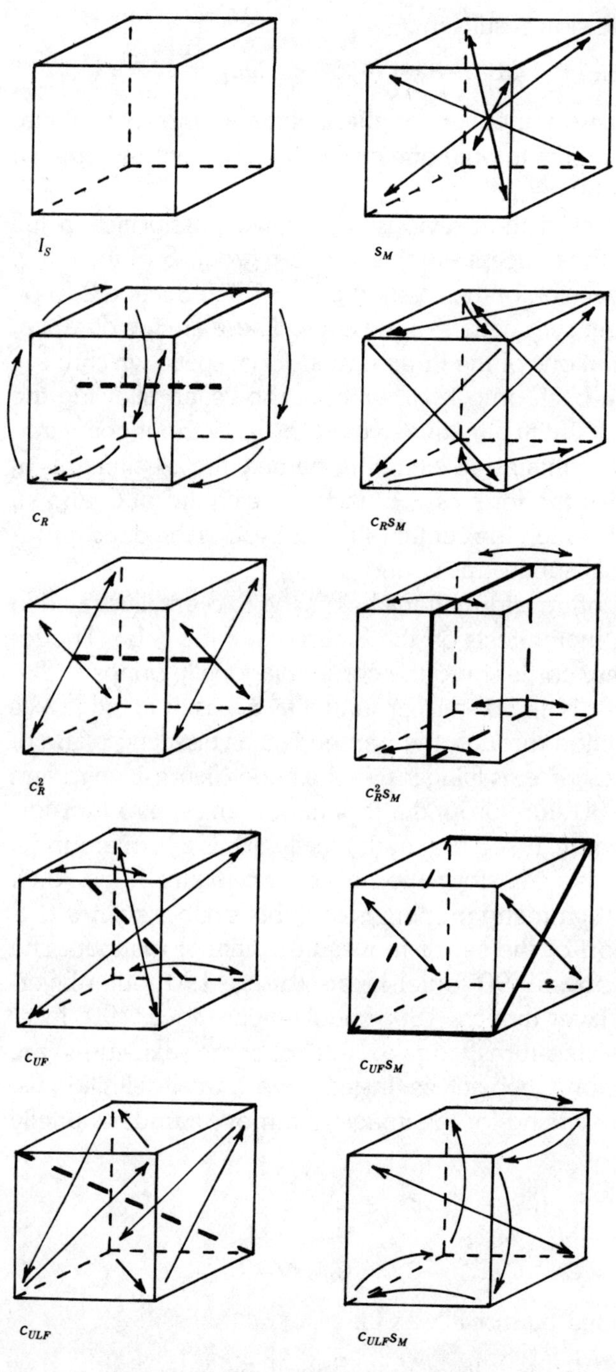

FIGURE 1.
The various types of elements in the symmetry group S of the cube

the bijections in the set of moves become bijections in the whole maneuver group $\overline{M}$ (or M when restricted to outer layer moves). For any arbitrary m, m_a, $m_b \in \overline{M}$ and s, s_a, $s_b \in S$ we have

$$m(s_a s_b) = (m s_a) s_b, \quad m I_S = m,$$
$$(m_a m_b) s = (m_a s)(m_b s).$$

The first two equations are summarized by the formulation "the group S acts on the *set* $\overline{M}$". According to the last equation, for every $s \in S$ the function defined on $\overline{M}$ is an endomorphism, and as a bijection therefore an automorphism. All three equations are summarized by the formulation "the group S acts on the *group* $\overline{M}$".

It is generally said a group B acts on a group A, if a function $A \times B \to A$, $(a, b) \to ab$, is defined which fulfills for any arbitrary a, $a_1, a_2 \in A$ and $b, b_1, b_2 \in B$ the equations $a(b_1 b_2) = (ab_1)b_2$, $aI_B = a$, $(a_1 a_2)b = (a_1 b)(a_2 b)$. We then always have $I_A b = I_A$ (since $I_A b = (I_A I_A)b = (I_A b)(I_A b)$) and $(ab)' = a'b$ (since $(a'b)(ab) = (a'a)b = I_A b = I_A$).

In the same sense, but at first independent of the maneuver group $\overline{M}$, S also acts on the group $\overline{G}$ of the possible operations, as f.i.

$$(uf, lb, rh)s_M = (db, rf, lf).$$

The connection with $\overline{M}$ is provided by the following equation valid for all $m \in \overline{M}$ and $s \in \overline{S}$:

$$(m\overline{\pi})s = (ms)\overline{\pi}.$$

In words, we obtain the same operation by two different methods: Either we first carry out a maneuver m and subsequently turn or mirror etc. the operation resulting from it. Or we first turn or mirror etc. the maneuver m, i.e. each one of its moves, and then carry out the maneuver resulting from it. Formulated differently again: From $m \to g$ we always get $ms \to gs$. Hence we get from

$$m_{520} = BM_R B^2 M_R' B \ (5) \to (uf, lb, rb)$$

and from the last two examples immediately

$$F' M_R F^2 M_R' F' \ (5) \to (db, rf, lf).$$

With this, we are finally prepared, to introduce the main concepts of this section:

Definition. (a) A maneuver m_a is called a *variation* of a maneuver m_b, if there exists an $s \in S$ with $m_a = m_b s$ or $m_a' = m_b s$. (b) An operation g_a is called a *variation* of an operation g_b, if there exists an $s \in S$ with $g_a = g_b s$ or $g_a' = g_b s$.

Theorem. The relation "is a variation of" is an equivalence relation in $\overline{M}$ or $\overline{G}$, respectively (similarly in M, G, G^* and $\overline{G}^*$).

Proof. It is sufficient, to prove the proposition for $\overline{M}$, since we exclusively use the abstract properties of the action of a group on a group deduced above. The relation is reflexive due to $m_a = m_a I_s$. It is symmetric, since from $m_a = m_b s$ there follows the equation $m_b = m_b I_s = m_b(ss') = (m_b s)s' = m_a s'$ and from $m'_a = m_b s$ first $m_a = m'_b s$ and then $m'_b = m_a s'$. To prove the transitivity we first assume $m_a = m_b s_a$ and $m_b = m_c s_b$ and obtain $m_a = (m_c s_b)s_a = m_c(s_b s_a)$. In the case $m_a = m_b s_a$ and $m'_b = m_c s_b$ we have $m'_a = m'_b s_a = m_c(s_b s_a)$. The two other cases are taken care of in the same manner.

The equivalence classes associated with this relation are called *variation classes* (of maneuvers or of operations). They consist of at most 96 different elements, since the symmetry group S has precisely 48 elements.

It is easy to find a maneuver, where all the 96 variations are different and cause 96 different operations, which are, of course, also variations of each other. A wonderful example of a maneuver with 96 different

5000 A.D.

variations which yet always cause the same operation, is–how could it be otherwise–our favorite maneuver

$$m_{490} = ((M_R U)^4 C_{ULF})^3 \ (24).$$

3.2 Combination

From the cyclic representation of the operation g caused by a maneuver m we can immediately read off the operation g^n caused by any arbitrary power m^n (n an integer). A suitable choice of n and the combination (meaning here a step by step execution, hence the composition) of different maneuvers often leads to the cancellation of individual unwanted cycles. Let us, for instance, investigate all two-move maneuvers with mutually perpendicular layers, of which there exist, variations excluded, only 11, as for instance RU, RU', RU^2, R^2U^2, M_RU, M_RU^2, RM_D^2, $R^2M_D^2$, M_RM_D, $M_RM_D^2$, $M_R^2M_D^2$. From $R^2U^2 \rightarrow$ (*ufl*, *ubr*, *dfr*)(*urf*, *drb*, *ulb*)(*ul*, *ur*, *dr*)(*uf*, *ub*)(*rf*, *rb*) we obtain the equally simple and useful maneuver

$$m_{620} = (R^2U^2)^3 \ (6) \rightarrow (uf,\ ub)(rf,\ rb).$$

$M_RU \rightarrow$ (*ufl*, *ulb*, *ubr*, *urf*)(+*uf*, *ru*)(+*ul*, *ub*, *bd*, *df*)(*u*, *b*, *d*, *f*) yields the important component

$$(M_RU)^4 \ (8) \rightarrow (+ul)(+ub)(+bd)(+df)$$

of our favorite maneuver m_{490} with the derivatives $(M_RU)^4(M_RU')^4$ (16) $\rightarrow$ (+*ur*)(+*ul*) and $(M_RU)^4(M_R'U)^4$ (16) $\rightarrow$ (+*uf*)(+ul)(+*ub*)(+*ur*). From $R^2M_D^2 \rightarrow$ (*urf*, *drb*)(*ubr*, *dfr*)(*ur*, *dr*)(*rf*, *lf*, *rb*, *lb*)(*l*, *r*)(*f*, *b*) we get

$$m_{600} = (R^2M_D^2)^2 \ (4) \rightarrow (fr,\ br)(fl,\ bl).$$

A brief digression: m_{600} and m_{620} transform the start position to positions with precisely four unicolored faces. It is easy to establish that there exist 33 positions of this kind, and with m_{600} (6 positions), m_{620} (12), $m_{600}C_Fm_{600}$ (3) and m_{625} (12) we obtain them all; the numbers in brackets here indicate the number of elements in the variation class of the operation. There is, of course, no position with precisely five unicolored faces. The two position induced by the maneuvers

$$D^2M_RM_FD^2M_F'M_R' \ (6) \text{ and } D^2(R^2B^2)^3D^2 \ (8)$$

from the start position show that five faces don't in general uniquely determine the sixth. We can also take I_Pm_{435} and produce a whole class

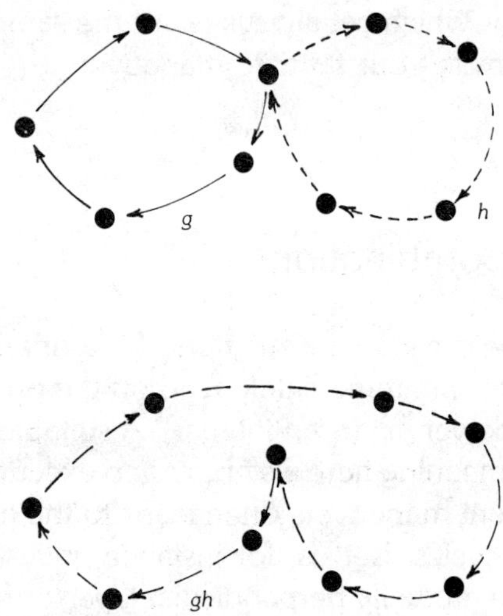

FIGURE 1.
Composition of two cycles with one common element

of not less than 12 positions with 5 identical faces by means of m_{510}, m_{610}, m_{615} etc.

As for the obvious question, how is the cycle structure of the composition of two operations determined by the cycle structure of the individual operations, we shall limit ourselves to a simple example: The composition of an r-cycle and an s-cycle ($r, s \in \mathbb{N}$), having precisely one element in common, is an $(r + s - 1)$-cycle (cf. figure 1). Thus we can, f.i., construct an edge-11-cycle by carrying out step by step two suitable edge-5-cycles and one suitable edge-3-cycle: We start with

$$m_{750} = D'M_R'DM_R\ (4) \rightarrow (dl, db, fd, uf, rd) \quad \text{(figure 2a)}.$$

Next, we carry out one of its variations,

$$RM_FR'M_F'\ (4) \rightarrow (rd, ur, lu, fr, rb) \quad \text{(figure 2b)},$$

and finally we conclude with a variation of our 4-move 3-cycle,

$$L^2M_DL^2M_D'\ (4) \rightarrow (rb, lb, lf).$$

The result is

$$D'M_R'DM_R \cdot RM_FR'M_F' \cdot L^2M_DL^2M_D'\ (12) \rightarrow (dl, db, fd, uf, ur, lu, fr, lb, lf, rb, rd).$$

By rearranging and condensing, the maneuver can be shortened to 10 moves.

Why doesn't the maneuver index list any of these wonderful edge-11-cycles? The answer is simple: There are too many of them. The "housing office" has 12 possibilities to pick 11 of the 12 edge cubies for the intended giant housing-ring-swap. Next, the 11 ready-to-move

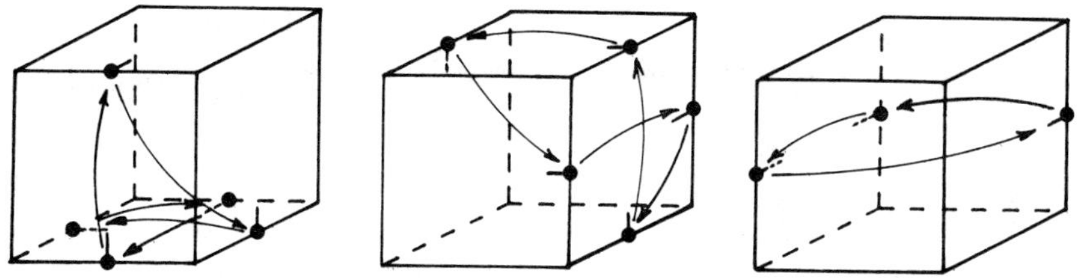

FIGURE 2. Components for the construction of an edge-11-cycle

cubies can be cyclically arranged by the "supervisory department" in 10! ways, and the "traffic department", finally, can orient them in 2^{10} ways, in brief, there are $12 \cdot 10! \cdot 2^{10} = 44\,590\,694\,400$ different possible edge-11-cycles. And since no variation class contains more than 96 elements we still have more than 464 million essentially different ones.

Following the same pattern we can compute the number of operations with an arbitrary given cycle structure. The cycle structure of an operation is here–in informal terms–the list of the numbers of all its cycles of different lengths and types. We denote the number of orientation-preserving or right-twisting or left-twisting corner cycles of length i by c_i or c_i^+ or c_i^-, respectively ($i = 1, \ldots, 8$), and the number of the orientation-preserving or reorienting edge cycles of length j by e_j or e_j^+, respectively ($j = 1, \ldots, 12$). In this way we can formally define the cycle structure of an operation as the 48-tuple $(c_1, \ldots, c_8, c_1^+, \ldots, c_8^+, c_1^-, \ldots, c_8^-, e_1, \ldots, e_{12}, e_1^+, \ldots, e_{12}^+)$. For enthusiasts –others should stay clear–here the desired generalization without proof:

Theorem. The number of operations with the cycle structure $(c_1, \ldots, c_8, c_1^+, \ldots, c_8^+, c_1^-, \ldots, c_8^-, e_1, \ldots, e_{12}, e_1^+, \ldots, e_{12}^+)$ is

$$\frac{8! \cdot 12! \cdot 3^{1\bar{c}_2 + 2\bar{c}_3 + \cdots + 7\bar{c}_8} \cdot 2^{1\bar{e}_2 + \cdots + 11\bar{e}_{12}}}{\bar{c}_1! 1^{\bar{c}_1} \cdot \bar{c}_2! 2^{\bar{c}_2} \cdot \ldots \cdot \bar{c}_8! 8^{\bar{c}_8} \cdot \bar{e}_1! 1^{\bar{e}_1} \cdot \bar{e}_2! 2^{\bar{e}_2} \cdot \ldots \cdot \bar{e}_{12}! 12^{\bar{e}_{12}}},$$

where $\bar{c}_i = c_i + c_i^+ + c_i^-$ and $\bar{e}_j = e_j + e_j^+$.

The theorem holds both for possible and impossible operations, or more precisely for "possible and impossible cycle structures" with $c_1 + 2\bar{c}_2 + 3\bar{c}_3 + \ldots = 8$ and $\bar{e}_1 + 2\bar{e}_2 + 3\bar{e}_3 + \ldots = 12$. Which cycle structures are possible or impossible can be determined with the help of the second law of cubology (Theorem 2.4.3). In the case of the edge-11-cycles, $c_1 = 8$, $e_1 = 1$ and $e_{11} = 1$, while all the other 45 components of the cycle structure are equal to zero.

3.3 Conjugation

This procedure was already used in the proof of the first law of cubology (Theorem 2.4.1). There, we started from a maneuver m for a special corner-3-cycle (X_1, X_2, X_3) and constructed a maneuver m^* for another corner-3-cycle, as f.i. (X_1, X_2, X_4). For this purpose we first transported the cubie from cubicle X_4 into X_3 by a small auxiliary maneuver m_a; next, we carried out m and finally cancelled out m_a again: $m^* = m_a m m_a'$ causes the corner-3-cycle (X_1, X_2, X_4). In the auxiliary maneuver m_a, only X_1 and X_2 must remain unchanged, all the other cubies need not concern us.

Here, the maneuver m is, in a sense, "tricked": Unaware that m_a has planted the changeling from X_4 among its wards, it brings up the three children (cycles them) and must witness later that the fat changeling is replaced again by the old hungry baby.

If a and b are elements of an arbitrary group A, we call b conjugate to a, if there exists an c in A with $b = cac'$. In this sense, m^* is conjugate to m in the maneuver group M. It causes an operation (X_1, X_2, X_4) which is conjugate to (X_1, X_2, X_3) in the group of operations (homomorphism $\bar{\pi}$).

The principle of conjugation is one of the most beautiful and powerful techniques for the construction of maneuvers. Since it is easy to bring three arbitrary given corner or edge cubies into three given corner or edge cubicles with given orientation–the rest of the cube may be "destroyed" at will–it f.i. permits to construct from *one* corner- or edge-3-cycle *any* arbitrary corner- or edge-3-cycle. Examples for conjugations which already appear in our "simple strategy" are the derivation of

$$m_{333} = FRUR'U'F' \text{ (6) from } RUR'U'$$

and of

$$m_{520} = BM_R B^2 M_R' B \text{ from } M_R B^2 M_R' B^2.$$

But also in the elementary maneuvers assembled in figure 1.3.1 for the construction of the first layer, conjugation is frequently used.

Because of the "prefix" m_a and the "suffix" m_a', to extend our grammatical lingo, the conjugated maneuver is in general longer than

its stem-maneuver. An important exception is the *shift*. It results from moving a part of the maneuver from one end to the other, in other words, from dividing a maneuver into two components and exchanging them. Indeed, for any arbitrary maneuvers m_a and m_b, the product $m_b m_a$ is conjugate to $m_a m_b$ since

$$m_b m_a = m_a' \cdot m_a m_b \cdot (m_a').$$

Serious cubologists should therefore first thoroughly shift every new maneuver. For this we have to carry it out only once if the cubies taking part in the operation we are interested in are previously marked in a suitable way (scotch tape, colored paper etc.). Example m_{520}, which we just mentioned, shows that it can be useful to divide all 180°-moves into two 90°-moves, and this in both ways (f.i. $R^2 = R \cdot R$ and $R^2 = R' \cdot R'$).

But back to our theory! It is easily seen that the relation "b is conjugate to a" defined in an arbitrary group is an equivalence relation and we can, therefore, speak of "elements conjugate to each other". (Reflexivity: $a = I_A a I_A$. Symmetry: From $b = cac'$ there follows $a = c'b(c')'$. Transitivity: From $b = d_1 a d_1'$ and $c = d_2 b d_2'$ there follows $c = (d_2 d_1) a (d_2 d_1)'$.) The equivalence classes associated with this relation are called *conjugacy classes* of the group.

We have seen that f.i. all corner-3-cycles belong to the same conjugacy class of Rubik's group G. Conversely, conjugate elements of the group G always have the same cycle structure since prefix and suffix result in only a temporary exchange of roles of the cubies. (For a formal proof note that for every $c \in G$ the map $g \to cgc'$ is an isomorphism of G onto G, a so-called "inner automorphism".) The set of all corner-3-cycles is therefore a conjugacy class of Rubik's group G (and of the group G as well).

Can we generally say that the set of operations with a given cycle structure is always a conjugacy class? The answer is no! If g is the product of a corner-7-cycle with an edge-11-cycle, as f.i. $g = (X_1, \ldots, X)(Y_1, \ldots, Y_{11})$, then g and $g^* = (X_1, \ldots, X_7)(Y_1, \ldots, Y_9, Y_{11}, Y_{10})$ have the same cycle structure but are not conjugate to each other.

3.4 Isolation

It is not hard to find a maneuver m which changes in a given layer c only one single cubicle c_o. As an example, the short maneuver $m = RDR'$ causes "isolated changes" in three layers at once: in the upper layer, in the U $-$ D middle layer and in the F $-$ B middle layer. If C is an arbitrary move with c, then in the maneuver $mCm'C'$ almost all the effects cancel themselves out. Only the cubie from c_o, its successor in cubicle c_o (with regard to the housing ring swap caused by m) and its predecessor in the new cubicle might not return home, due to the "intermediate move" C. These are at most three cubies and we obtain either a 3-cycle or the twisting of two corner cubies or the reorientation of two edge cubies.

Example 1. $RDR' \cdot U \cdot RD'R' \cdot U'$ (8) $\rightarrow$ (*ubr*, *ulb*, *rbd*).
This maneuver for a corner-3-cycle is a variation of m_{104}.

Example 2. $RDR' \cdot M_D' \cdot RD'R' \cdot M_D$ (8) $\rightarrow$ (*df*, *br*, *lb*).
This maneuver for an edge-3-cycle is a variation of m_{521}.

The situation becomes particularly clear if the isolated change in the cubicle c_o consists of a turning of the cubie in the same spot. Here we talk of an *isotwist* (isolated twist of a corner cubie) or an *isoflip* (isolated reorientation of an edge cubie).

Example 3. (D. Goto). $m_o = R'DRFDF'$ (6) $\rightarrow$ (+*urf*) $\cdot \ldots$
causes an isotwist for the upper layer, from which we get

$$\begin{aligned} m_{5a} &= m_o U' m_o' U \\ &= R'DRFDF' \cdot U' \cdot FD'F'R'D'R \cdot U \ (14) \rightarrow (+urf)(-ufl) \end{aligned}$$

as well as $m_o U m_o' U'$ (14) $\rightarrow$ $(+urf)(-ubr)$ and $m_o U^2 m_o' U^2$ (14) $\rightarrow$ $(+urf)(-ulb)$.

Example 4. (D. Seal) $m_{400} = RM_DR^2M_D^2R$ (5) $\rightarrow$ $(+ur) \cdot \ . \ . \ .$ causes an isoflip for the upper layer from which we get

$$\begin{aligned} m_{415} &= m_{400}U'm'_{400}U \\ &= RM_DR^2M_D^2R \cdot U' \cdot R'M_D^2R^2M'_DR' \cdot U \ (12) \rightarrow (+uf)(+ur) \end{aligned}$$

as well as $m_{400}U^2m'_{400}U^2$ (12) $\rightarrow$ $(+uf)(+ub)$.

Example 5. (F. Barnes) $m_{401} = R'FD'RF'$ (5) $\rightarrow$ $(+fr) \cdot \ . \ . \ .$ causes an isoflip for the $U - D$ middle layer from which we get

$$\begin{aligned} m_{410_a} &= m_{401}M_D^2m'_{401}M_D^2 \\ &= R'FD'RF' \cdot M_D^2 \cdot FR'DF'R \cdot M_D^2 \ (12) \rightarrow (+fr)(+bl). \end{aligned}$$

With $m_{405_a} = m_{401}M'_Dm'_{401}M_D$ (12) $\rightarrow$ $(+fr)(+br)$ we obtain a variation of the operation last considered in Example 4.

One remaining flaw is that the last three iso-maneuvers, m_o, m_{400} and m_{401}, must be memorized "forward" and "backward". This drawback can be avoided with the help of isotwists of order 3 and isoflips of order 2 with somewhat longer maneuvers. They are always used "forward", making use of the fact that we always need a number of right twists divisible by 3 (f.i. one right turn and one left turn) and an even number of flips.

Example 6. $m_1 = (RF'R'F)^2$ (8) $\rightarrow$ $(+urf) \cdot (+dfr)(-dlf)(-drb)(fr, dr, fd)$ causes an isotwist of order 3 for the upper layer. Possibilities of application are f.i. $m_1U'm_1^2U$ (26) $\rightarrow$ $(+urf)(-ufl)$ and $m_1Um_1U^2m_1U$ (27) $\rightarrow$ $(+ufl)(+urf)(+ubr)$.

Example 7. $m_{402} = (M_DR)^4$ (8) $\rightarrow$ $(+ur) \cdot (+fl)(+lb)(+br)$ is an extremely simple maneuver for an isoflip of order 2 both for the upper layer and for the $F - B$ middle layer. It is particularly versatile in its applications, as f.i. with

$$m_{402}U'm_{402}U \ (18) \rightarrow (+uf)(+ur) \text{ and } (m_{402}M_F^2)^2 \ (18) \rightarrow (+ur)(+dl).$$

Maneuvers constructed according to the isolation principle have the form of a commutator in the maneuver group, and hence also cause (via the homomorphism) a commutator in the group of operations. But the frequent appearance of commutators in "good" maneuvers also has to do with the fact that our remarks about the special commutators $mCm'C'$ at the beginning of this section are more generally valid. If m_a and m_b are arbitrary maneuvers and C_a and C_b are the set of cubicles inside of which m_a respectively m_b produce a change, then in $m_am_b m'_am'_b$ the results of the four individual maneuvers cancel each other out with the exception of the movements of those cubies that enter the "turn-table" $C_a \cap C_b$ at a critical moment. If, for example, $m_a = R$ and $m_b = U$, we have $C_1 \cap C_b = \{urf, ur, ubr\}$, and the changes caused by $RUR'U'$ are limited to the seven cubies lying on the three cube edges ub, ur and fr.

Example 8. (J. Conway) $FDF^2D^2F^2D'F'$ (7) $\rightarrow$ $(ufl, fur) \cdot \ldots$
causes an "isoswap" (isolated swap of two cubies) for the upper layer. Similarly, we have

Example 9. $m_{300} = R'DF'D^2FD'R$ (7) $\rightarrow$ $(ufl, urf) \cdot \ldots$,
but this isoswap is even "strictly orientation-preserving" with regard to the upper cube face. Both these operations are of the order 2. This can be seen without explicit execution, since already the maneuvers in the maneuver group are of the order 2 (inverses to themselves) and obviously do not cause the identity. With m_{300} we obtain for example

$$m_{210} = (m_{300}U^2)^2 = (R'DF'D^2FD'RU^2)^2 \rightarrow (ufl, urf)(ulb, ubr).$$

It is fascinating and instructive to see how many old acquaintances are modelled after this pattern. Thus, $m_{500} = R^2M_D'R^2M_D$ (4) $\rightarrow$ (fr, bl, br) and $m_{600} = (R^2M_D^2)^2$ (4) $\rightarrow$ $(fr, br)(fl, bl)$ make use of the isoswap (fr, br) caused by R^2 on the $U - D$ middle layer, whereas $m_{898} = M_R^2M_D'M_R^2M_D$ (4) $\rightarrow$ $(f, b)(l, r)$ employs the isoswap (f, b) caused by M_R^2 on the same middle layer.

4 PRETTY PATTERNS

"I can't stand the creaking of that cube any longer!" Every reader sharing his cubicle–or rather his domicile–with somebody not sharing his fascination, will, at this point of his studies, be familiar with shrieks or groans of that kind. However, the unfortunate fellow lodger, unable to enjoy either the sophisticated mechanics of the cube or its intricate logic, might be placated by the visual aesthetics, the optical and artistic attraction of a pretty pattern.

Many new patterns or new maneuvers for already known patterns have been contributed by Peter Klerings (Bonn) and Rainer aus dem Spring (Uberhausen), who will be referred to in this Chapter simply by their initials PK and a.d.S.

4.1 Christmas Parcels and Chicken Feet

At least three relatively simple patterns are quickly discovered by many puzzlers; they are the **Four Dots,** produced from the start position by

$$p_1 = m_{898} = M_R^2 M_D' M_R^2 M_D \ (4),$$

the **Six Dots,** which we can also reach in four moves by

$$p_2 = m_{899} = M_R M_F' M_R' M_F \ (4) \quad \text{(picture 3 left)},$$

and the **Chessboard Cube of Order 2,** caused by a 180°-turn of the three middle layers in any arbitrary order, as f.i.

$$p_3 = m_{790} = M_R^2 M_F^2 M_D^2 \ (3) \quad \text{(picture 3 right)}.$$

However, there exist precisely 17 positions with each of the six cube faces showing a two-colored chessboard pattern. Apart from the one already mentioned there are the 8 **Chessboard Cubes of Order 3** of the type

$$p_4 = m_F U' M_R' B' M_R^2 B M_R' D \cdot U M_R' F M_R^2 F' M_R' D' M_F' \ (16)$$

(picture 4 left, P. Klerings) and the 8 **Chessboard Cubes of Order 6** of the type

$$p_5 = p_4 p_3 \quad \text{(picture 4 right)}.$$

We have started with this example, because it probably serves best to demonstrate, that classifying positions according to the type of the six cube faces must necessarily remain a "surface" enterprise, in every sense of the word. Yet, it has its justifications, since it is the surface of the cube which makes up the beauty of a position. But for a deeper understanding of the structure, we need a different approach: Many of the positions we consider beautiful are produced from the start position by applying an element of the symmetry group S of the cube (cf. Section 3.1) to a subsystem of the cube distinguished by symmetry properties, while the rest of the cube remains fixed. Occasionally, different elements of the group S operate on different subsystems. In the case of the Chessboard Cubes the distinguished subsystem is the set of the 12 edge cubies. It is mirrored through the center of the cube (reflection S_M) to produce the Chessboard Cube of Order 2. In the case of the Chessboard Cubes of Order 3 the set is turned by 120° around a corner axis. (The maneuver p_4 causes the corner axis turn c_{URF}.) 4 corner axes, each with 2 rotation possibilities, make up the 8 positions of this type. In the third case (Chessboard Cube of Order 6) the edge cubie system is subjected to a rotational reflection of the type $c_{URF}s_M$. Here too, there exist 8 positions, since there is a one-to-one relation between the rotational reflections of this type and the corner axis turns.

A look at the notation for the operations of the different cube moves given at the end of Section 1.5 shows, that the edge cubie system does not allow isolated edge axis turns and 90° face axis turns. The former require two edge flips and five transpositions, the latter three 4-cycles; in other words, they both require odd permutations. However, isolated

180° face axis turns with six transpositions are possible. They produce **Four Chessboard Faces,** in the case of c_U^2 f.i. by

$$P_6 = R^2F^2M_R^2B^2L^2M_D^2C_F^2 \ (6).$$

Followed by s_M the edge cubie system as a whole has been simply mirrored through the horizontal plane ($c_U^2 s_M$, cf. figure 3.1.1). We end up with a position with **Two Chessboard Faces** (and four unicoloured faces) which can also be produced directly by means of

$$p_7 = F^2R^2M_F^2L^2F^2M_R^2 \cdot C_U^2 \ (6).$$

Already for mechanical reasons we cannot subject the rigid face cubie system or any arbitrary corner cubie to the point reflection s_M. As soon as our "distinguished subsystem" contains all the face cubies or at least one corner cubie, only the so-called proper movements, i.e. the elements of the rotation group C, need concern us. Our first example for this is the ring system consisting of all the edge and face cubies, or–dually–the system of the 8 corner cubies. The easiest here is an edge axis turn, f.i. c_{UB} for the corner cubie system by means of

$$p_8 = R'm_{350}R = R'M_R^2M_F^2U^2M_R^2M_F^2D^2R \ (8)$$

(**Christmas Parcel of Order 2,** picture 5 left).
Due to

$$c_{UB}c_{UL} = c_{UFR}$$

two edge axis turns lead to a corner axis turn and we end up with the pretty **Christmas Parcel of Order 3** (picture 5 right)

$$p_9 = p_8C_Up_8C_U' \ (16).$$

For the 6 edge axis turns we have 6 corresponding Christmas Parcels of Order 2, for the 8 corner axis turns we have 8 Christmas Parcels of Order 3, and these 14 positions make up all the "Six Plus" patterns.

However, the system of the 8 corner cubies can also be turned around a face axis: by 180° using

$$p_{10} = m_{350} = M_R^2M_F^2U^2M_R^2M_F^2D^2 \ (6),$$

and by 90° using

$$p_{11} = m_{365} = U^2M_R^2UM_R^2U^2 \cdot M_F^2D'M_F^2 \ (8) \quad \text{(picture 6 left)},$$

where in both cases we end up with **Four Plus** and two unicolored faces. Since every corner axis turn is also the product of two 90° face axis turns, f.i.

$$c_Bc_U' = c_{UFR},$$

the Christmas Parcel of Order 3 can also be consructed by means of p_{11}:

$$p_{9a} = C'_R p_{11} C_R p'_{11}.$$

By means of skillful rearrangements two moves cancel themselves out and we get

$$p_{9b} = BM_R^2 B^2 M_D^2 F' M_D^2 B^2 \cdot U^2 M_F^2 D M_F^2 U^2 M_R^2 U' \ (14) \quad \text{(H. Kraß)}.$$

At this point, we have complete command over all the isolated possible movements of the total system of the face cubies (p_1 and p_2), of the total system of the edge cubies (p_3 to p_7) and of the total system of the corner cubies (p_8 to p_{11}), and are now able to combine these three types of movements any way we want. A particularly attractive pattern with **Six Plus-Dot-Faces** (picture 6 middle) results from turning edge cubie system and face cubie system around the same corner axis but in different directions:

$$p_{12} = p_{9b} p_2 \ (18).$$

Other pretty combinations are

$$p_{13} = p_{9b} p_3 \ (17)$$

which also produces **Six Plus-Dot-Faces,** and

$$p_{14} = p_3 p_2 \sim M_R^2 M_D M_F^2 M_D \ (4)$$

with **Four Plus and Two Chessboard Faces** (picture 6 right).

A quite different distinguished subsystem consists of two diametri-

cally opposed "peaks", with each peak consisting of a corner cubie and its three neighboring edge cubies. By means of

$$p_{15} = L^2UF^2L^2D'L'R'DM_RD'RDM_R'L'FU'FL^2 \ (18) \quad \text{(P. Klerings)}$$

the peaks at urf and dbl are subjected to the corner axis turn c_{URF}: **Twisted Peaks** (picture 7 left). Similar to the Six Dots, the Chessboard of Order 3 and the Christmas Parcel of Order 3 there exist 8 positions of this kind. By means of

$$p_{16} = U^2R'D'BRM_D^2R'B'DRF^2LFU'FD^2F'UF \cdot C_U^2 \ (19) \quad \text{(a.d.S.)}$$

the same peaks are exchanged by the edge axis turn c_{UB}: **Exchanged Peaks** (picture 8 left). Since each of the 4 pairs of peaks can be exchanged in 3 different ways we get 12 positions.

If in each of the two peaks the corner cubie is replaced by its three neighboring face cubies, a new distinguished subsystem is produced: two diametrically opposed rings permitting the same movements as the peaks. The maneuver

$$p_{17} = m_{758}p_2 \sim M_D'M_R'M_DF'R^2DB'M_R^2BM_R^2D'R^2FM_R \ (14) \quad \text{(PK)}$$

causes the corner axis turn C_{URF} of the rings around *urf* and *dbl*: **Twisted Rings** (picture 7 right). **Exchanged Rings** (edge axis rotation C_{UB}, picture 8 right) are produced by

$$p_{18} = UM_R'U'B^2UM_RURM_F^2L'B^2D^2 \cdot C_R' \ (12) \quad \text{(a.d.S.).}$$

As in the case of the peaks, there exist 8 positions of the type of the Twisted Rings and 12 positions of the type of the Exchanged Rings.

Attractive modifications of the twisted (and, therefore, often also of the exchanged) rings and peaks are produced, if one of the two edge-3-cycles is omitted or replaced by its inverse, if (in the case of the rings) edge and face cubies at one or at both corners are rotated in opposite directions, or–particularly ingenious–if the two rotation symmetrical edge-3-cycles are replaced by *reorienting* edge-3-cycles which are also rotation symmetrical. We get, among others,

$p_{19} = m_{550}p_2, \quad p_{20} = m_{550}p_2',$

$p_{21} = m_{759}p_2, \quad p_{22} = p_{21}p_{18},$

$p_{23} = p_2'm_{758}$ (**Gay Rings,** picture 17 left), $p_{24} = p_{23}p_{18},$

$p_{25} = m_{440}, \quad p_{26} = p_{25}p_{18},$

$p_{27} = m_{440}p_2, \quad p_{28} = p_{27}p_{18},$

$p_{29} = m_{440}p_{17}, \quad p_{30} = p_{29}p_{18},$

$p_{31} = m_{440}p_{21}, \quad p_{32} = p_{30}p_{18},$

$p_{33} = m_{440}m_{758}m_{20}'$ (**Little Windmills**),

$p_{34} = p_{32}p_2'$ (**Big Windmills**).

Together with its three adjoining face cubies each corner cubie forms a "chicken foot". Chicken feet are treated like peaks and rings:

$p_{35} = m_{20}p_2 = (R'D^2RC_{FL})^4M_RM'_FM'_RM_F$ (16)
(Twisted Chicken Feet, picture 9 left),

$p_{36} = U^2B^2RBR'BU'M_F^2M'_DM'_FD^2R^2DM'_RD'R^2DR'UL \cdot C_{ULF}$ (20)
(Exchanged Chicken Feet, picture 10 left, a.d.S.).

Somewhat larger than the chicken foot is the "duck foot". It consists of three edge and three corner cubies, that is of all the cubies sitting on the three edges of the cube which meet in one fixed corner, with the exception of the central corner cubie sitting on all three edges. Two diametrically opposed duck feet can rotate in the same or the opposite direction and exchange their place by an edge axis rotation; m_{921} turns a single duck foot.

$p_{37} = m_{921} = U'R^2D'B'DR'D'BDR'U$ (11) **(Twisted Duck Foot),**
$p_{38} = (p_{37}C_{UB})^2$ (22) **(Twisted Duck Feet 1,** picture 9 right),
$p_{39} = p_{37}C_{UB}p'_{37}$ **(Twisted Duck Feet 2),**
$p_{40} = RB^2RLB^2R^2L'D'RL^2DFD'L^2UB'UBU^2R'D$ (21)
(Exchanged Duck Feet, picture 10 right, a.d.S.).

From the twisted peaks, rings and feet it isn't far to one of the most impressive patterns, the **Cubes within the Cube** (picture 11 left). It consists of two diametrically opposed $2 \times 2 \times 2$-cubes which intersect in the fictitious central cubie (space cubie) and rotate around the connecting corner axis by means of

$p_{41} = RF^2R'U'R^2U^2R'UFDBR'B'D'U^2F'RF^2R' \cdot C_{URF}$ (19) (PK).

However, the two cubes can only be exchanged by breaking forcibly into another orbit: seven transpositions amount to an odd permutation, hence an impossible operation. Closely related to the Cubes within the Cube are

$p_{42} = RF^2R'U'R^2U^2R^2UFU^2F'RF^2R'$ (14)
(Twisted Cube Edges, PK) and the strange crossbreed

$p_{43} = D^2FR^2D^2F'M_RM'_FL'DR'BU'BR^2 \cdot C_{UFL}$ (14)
(Cube within the Cube and Chicken Foot, PK).
a.d.S.

Two $2 \times 2 \times 2$-subcubes which are not situated opposite each other on a space diagonal but merely on a plane diagonal, can be exchanged by means of a 180° face axis rotation (Ronald. L. Fletterman and a.d.S.):

$p_{44} = D^2F'D^2R^2FRF'RU^2L'BLM_D^2B$ (14)
("Ron's Cubes within the Cube").

We end this section with a few simple pretty patterns for relaxation:

$p_{45} = (FBRL)^3$ (12) (**Four Diagonals,** picture 12 left).
$p_{46} = (F'B'RL)^3$ (12) (**Four Diagonals and Two Plus,** picture 12 middle).
$p_{47} = p_{46}M_D^2$ (13) (**Four Z and Two Plus**).
$p_{48} = p_2 m_{126} \cdot (dbl)$ ("**Six Diagonals,** picture 12 right).

The latter is one more example of a pretty pattern in another orbit. It is easy to show (systematic case distinction), that all patterns with six pure diagonals are impossible. However, we can get **Six Diagonals** on 3-colored faces by rotating four peaks, f.i.

$$p_{49} = RLUR'L'UF'B'D'FB \cdot RLD'R'L'U'F'B'UFB \ (22) \quad (\mathrm{PK}).$$

4.2 Snakes

If we rotate the Cubes within the Cube pattern around the corner axis connecting the two $2 \times 2 \times 2$-cubes we get a zigzag pattern meandering around the entire cube: We've reached the realm of the snakes! Its two most prominent specimens can be charmed like this:

$$p_{60} = m_{762}p_2$$
$$\sim R'B^2D'L'M_FLM_D'M_FUR^2F \cdot C_U' \ (11) \quad (\mathrm{a.d.S.})$$
$$\sim D'F^2R' \cdot M_FU^2DM_RU^2D' \cdot RF^2D \ (12) \quad (\mathrm{PK})$$
$$p_{61} = UM_F'U'F^2 \cdot UM_FU'F^2 \cdot LU^2M_R'D^2R' \cdot C_R \ (13)$$
(**Python,** picture 13 right, R. Walker).

The Anacondas are corner axis rotators and appear, therefore, in 8 species, while the Python family belongs to the edge axis rotators and is–thanks to a strange asymmetry–divided into two genera of 6 species each.

To study the anatomy of the Anaconda we cut a cube with the perpendicular bisecting plane of two diametrically opposed corners, as f.i. urf and dbl. This plane bisects 6 of the 12 cube edges and the 6 points of intersection are the corners of a regular hexagon, the "edge hexagon" (figure 1). The corresponding edge cubies form another distinguished subsystem which is of interest for its own sake. It can be turned by 120° around the corner axis perpendicular to the hexagon (two 3-cycles), in our example c_{URF} by

$$p_{62} = m_{762} = D'F^2R' \cdot M_FD^2UM_F'D^2U' \cdot RF^2D \ (12) \quad (\mathrm{PK})$$

or

$$p_{62a} = U'R'F^2RF \cdot URU^2R'F' \cdot F'U'R^2UR \cdot FUF^2U'R' \ (19) \quad (\mathrm{R.\ Walker})$$

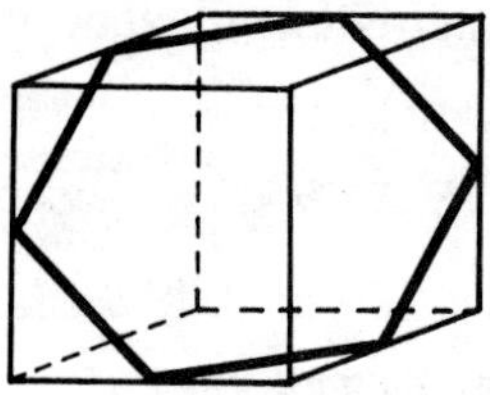

FIGURE 1. The edge hexagon

(**Edge Hexagon of Order 3,** picture 14 left, 8 positions), and by 180° around each of the three edge axes which connect two opposite corners of the hexagon, as for example c_{UB} by

$$p_{63} = RU^2F^2U^2R^2M_RUM_RU'F^2UM'_RUM'_RF^2R \ (16)$$
(**Edge Hexagon of Order 2,** picture 14 right, 12 positions, P. Klerings).

Obviously, our Anaconda is nothing but the product of an Edge Hexagon of Order 3 with Six Dots: $p_{60} \sim p_{62}p_2$. Its construction corresponds exactly to that of the "Twisted Rings" out of two smaller edge-3-cycles and p_2 already practiced in Section 4.1. There we got a whole family of pretty ring patterns by inverting one or both edge-3-cycles, by proceeding from orientation-preserving edge-3-cycles to reorienting edge-3-cycles, and by exchanging the rings via an edge axis turn. Similarly we can now breed some new colorful snake species:

$p_{64} = p_{62}p'_2$ (**Anaconda multicolor 1,** picture 17 right),
$p_{65} = p_{60}m_{441}$ (**Anaconda multicolor 2),**
$p_{66} = p_{64}m_{441}$ (**Anaconda multicolor 3).**

Rainer aus dem Spring discovered six "open" snakes. While the first two are distant relatives of our corner axis rotating Anaconda, the last four have more biogenetic relationships to the edge axis rotating Python:

$p_{67} = F'LDM^2_FUM_FD'M'_FU'M_FL'F'R'M_DRF^2$ (16) **(Black Mamba),**
$p_{68} = M'_RULM'_FL^2RUM_RU^2M'_RUM_RB'M_RUM'_RU' \cdot C'_R$ (17) **(Green Mamba)**
$p_{69} = R^2L'U^2RLU^2RULF^2L'FD^2F'LF^2L'U'$ (18) **(Female Rattlesnake)**
$p_{70} = p'_{69}$ **(Male Rattlesnake)**
$p_{71} = R^2LF^2M^2_RFU'LF^2M^2_DF^2M^2_DL'UF'B^2L \cdot C^2_R$ (16) **(Female Boa)**
$p_{72} = p'_{71}$ **(Male Boa)**

Very few snakes practice active brood care. One of them is

$p_{73} = UR \cdot F'M^2_DFBM_RB' \cdot R'U'$ (10) **(Cobra with Baby Cobra).**

We, therefore, better leave the realm of the snakes and turn to a somewhat more pleasant theme:

4.3 Birthday Cubes

Admittedly, in our electronic age, there are more economical information storages than Magic Cubes. However, cubologists called Therese, Tina, Theodor, Tom etc. might be pleased to find a Magic Cube with the letter T on each of its six faces on their birthday table. Tom Werneck, a well-known German game-expert and cube-author, created a complete "cube-alphabet". For the meticulous purist–and pedantry is an occupational disease mathematicians are prone to–the letters *H, I, L, O, S, T, U, X, Y* and *Z* are particularly attractive:

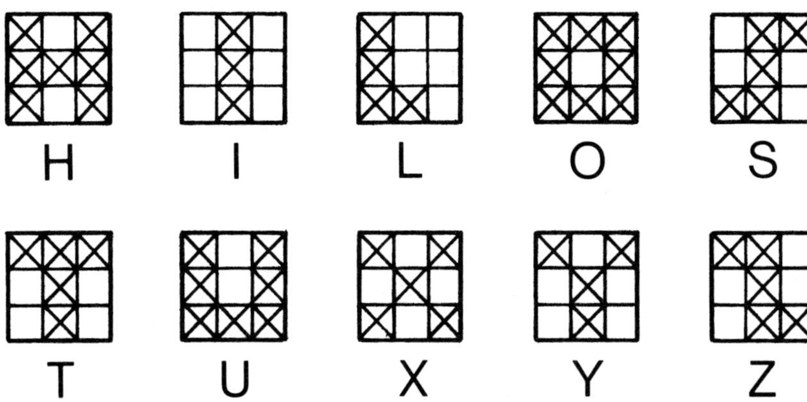

FIGURE 1. Cube letters

With *O* and *X* we have already dealt in Section 4.1 under the names "Dot" and "Chessboard Face". Here we give some further examples, many of which are again due to R. aus dem Spring:

$p_{80} = m_{773} = D^2M_RM_F^2M_R'U^2 \cdot C_U^2$ (5) (**Six H,** type 1),
$p_{81} = m_{791} = F^2R^2U^2M_F^2D^2R^2F^2 \cdot C_F^2$ (7) (**Six H,** type 2),
$p_{82} = M_R^2FL^2M_D^2R^2M_D^2B' \cdot C_R^2$ (7) (**Six H,** type 3),
$p_{83} = U'M_R^2F^2M_R^2B^2D'M_F^2D^2 \cdot C_F^2$ (8) (**Six H,** type 4),
$p_{84} = U^2L^2F^2UM_F^2U'M_R^2B^2L^2U^2 \cdot C_U^2$ (10) (**Six H,** type 5),
*p_{84} and p_{91} follow immediately one after the other!
$p_{91} = DLF' \cdot D'M_RDUM_F^2U' \cdot FL'D'$ (12) (**Six U,** type 1),
$p_{92} = RM_DR^2D'R'M_DF'U'LUFM_D'RU \cdot C_{UBL}$ (14) (**Six U,** type 2),
$p_{93} = p_{92}'$ (14) (**Six U,** type 3),
$p_{94} = U' \cdot FR'DM_RD'RF'RUM_F^2U'R' \cdot U$ (14) (**Six U,** type 4),
$p_{95} = p_{94}'$ (14) (**Six U,** type 5),
$p_{96} = M_D'FU'M_R'UF'D'M_FU \cdot C_U'$ (9) (**Six U,** type 6, picture 15 left),

$p_{97} = R'U'M_FURM'_DRUM_FU'R'M_D$ (12) (**Six U,** type 7),
$p_{98} = M_FUM_RU^2B^2D'M_FDBFR \cdot C'_F$ (11) (**Six U,** type 8),
$p_{99} = R' \cdot DUBUM_FU'B'D'U'F'U'M_RUF \cdot R$ (16) (**Six U,** type 9).

The following Six U pattern contains our personal specialty, two reorienting rotation symmetrical edge-3-cycles. We, therefore, change its name ruthlessly to "Six C" (C standing for Christoph):

$p_{100} = R'U' \cdot BRFM_RB'D'FM'_DF^2 \cdot UR$ (13)
(**Six C,** picture 15 right).
$p_{101} = M_R^2UM_R^2U^2M_R^2DM_F^2M_D$ (8) (**Four U,** type 1),
$p_{102} = DRFM_RDRB'R'D'M'_RF'R'$ (12) (**Four U,** type 2),
$p_{103} = M'_RD^2M_RU^2(F^2D^2)^2$ (8) (**Six Little U,** type 1),
$p_{104} = FBD^2M_F(U^2R^2)^2U^2F^2$ (10) (**Six Little U,** type 2),
$p_{105} = U^2F^2R^2B^2D^2F^2D^2R^2D^2F^2$ (10)
(**Six T,** picture 16 left),
$p_{106} = M_R^2D^2(M_R^2D)^2UM_F^2U$ (9) (**Four T,** type 1),
$p_{107} = RBM_RURFR'U'M'_RB'R'$ (11) (**Four T,** type 2)
$p_{108} = LU^2B^2U^2F^2RF^2U^2B^2L'R^2U^2LR^2$ (14) (**Six Little T**),
$p_{109} = DM_R^2U^2M_R^2D$ (5) (**Four Little T**)
$p_{110} = RL \cdot UD \cdot R'L' \cdot F'B' \cdot UDFB$ (12)
(**Six Double L,** picture 16 right),
$p_{111} = UDFBM_RF'B'$ (7) (**Four Double L and Two O**),
$p_{112} = R^2F^2M_R^2B^2L^2$ (5) (Four I, type 1),
$p_{113} = R^2D^2M_R^2U^2R^2$ (5) (Four I, type 2),
$p_{114} = p_{112}M_R = R^2F^2M_R^2B^2L^2M_R$ (Six I),
$p_{115} = UDM_F^2U'D'M'_R$ (6) (**Four I and Two O**)

What does a cubologist do, when he finds himself shipwrecked on a deserted island, of course still clinging to his cube? He estimates the current of the sea around him, works on his cube with

$p_{116} = R^2B^2L^2D^2 \cdot M_R^2B^2M'_RB^2 \cdot U(R^2U^2)^2R^2U \cdot F^2M_DF^2M'_D$ (19)
(**Rescue Cube,** picture 18 right)

producing a double SOS-call and looks forward to his next birthday which will be celebrated by streaming banners:

$p_{117} = F'DR^2D'RB^2R'DR^2M_DF^2U'FL^2F'UF^2U'R$ (19)
(**Four Tricolours**), a.d.S.
$p_{118} = F'DR^2D'RB^2R'DR^2M_DF^2U'FL^2F'UF^2U'L'$ (19)
(**Six Tricolours,** type 1), a.d.S.
$p_{119} = R^2LB'DL^2D'LF^2L'DL^2M_DB^2U'BR^2B'UB^2U'LF$ (22)
(**Six Tricolours,** type 2, picture 18 left). a.d.S.

5 VARIATIONS AND GENERALIZATIONS

By means of a slight modification every Magic cube can be transformed into a "Supercube", a puzzle with over 2000 times as many positions. In the first section of this chapter we give a solution for the supercube and a suitable mathematical model. In the second, we pay our respects to an over one hundred year old predecessor of the Rubik's cube. We present a possibly new and certainly very simple proof of the main theorem of Sam Loyd's 15-Puzzle. Most of the objects treated in the third section are no longer cube-shaped, but due to a technical and structural kinship, they are all part of our science of cubology: We are talking of magic polyhedrons of all kinds.

5.1 The Supercube

In this section cubology turns into a genuinely occult science: Its operations are no longer visible!

Let us, for example, look at the 2-move maneuvers

$$UR \rightarrow (-ufl,\ ulb,\ bdr,\ dfr,\ fur)(+ubr)(uf,\ ul,\ ub,\ br,\ dr,\ fr,\ ur),$$
$$UR' \rightarrow (-ufl,\ ulb,\ fur)(+ubr,\ frd,\ drb)(uf,\ ul,\ ub,\ fr,\ dr,\ br,\ ur).$$

From the cyclic structure of their operations we find

$$(UR)^{105}(210) \sim I_M,$$
$$(UR')^{63}(126) \sim I_M.$$

Hence, the maneuvers $(UR)^{105}$ and $(UR')^{63}$ bring every cube from the initial state back into the initial state. But they do, all the same, have an effect: Since during the first maneuver the upper and the right outer layer are turned clockwise $105 = 26 \cdot 4 + 1$ times, we find that the face cubies u and r are turned by 90° to the right after the maneuver. Analogously in the second maneuver, u is turned by 90° to the left and r by 90° to the right. For this we briefly write:

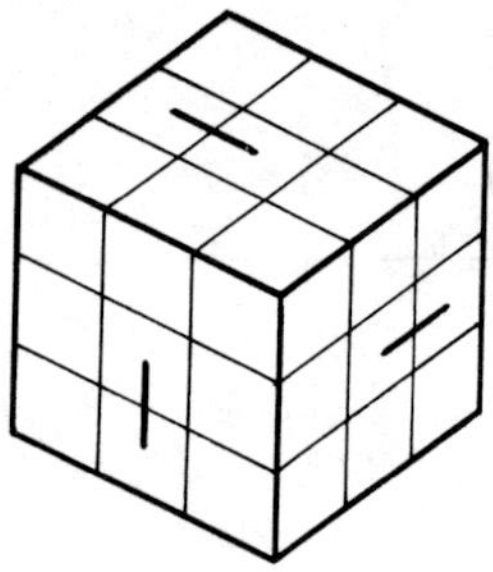

FIGURE 1. A Supercube

$$(UR)^{105}(210) \to (+u)(+r),$$
$$(UR')^{63}(126) \to (-u)(+r).$$

The 180° turn of a face cubie is denoted by two plus signs put side by side, f.i.

$$(UR)^{210}(420) \sim (UR')^{126}(252) \to (++u)(++r).$$

There are several ways to visualize the normally invisible turns of the face cubies: different coloring, pretty patterns, or ugly commercials. But the simplest method seems to mark each of the six cube faces by a line connecting the center of the colored tile of the face cubie with one of the neighboring tiles. In figure 1 the lines are arranged in such a way that on each of the 12 cubies forming an anaconda there lies half a line (cf. chapter 4).

The goal of the "supercube" game is to "re-turn" a marked and scrambled cube back to its start position in such a way that also the line segments match. If we simply disregard the lines, our chances of accidentally stumbling on the start position of the supercube by reaching the start position of the ordinary cube are rather slim—1:2048 to be exact. In fact, it follows already from the two maneuvers discussed above that in every position (in the sense of the ordinary cube) the face cubies can have at least $4^5 \cdot 2 = 2^{11} = 2048$ different combinations of positions: Five of the six face cubies can be turned independently of one another into each of four different positions, and for the sixth cubie there still exist two possibilities. That there are no more follows from theorem 1, which will be proved later.

Hence, the two beautiful but lengthy maneuvers $(UR)^{105}$ and $(UR')^{63}$ together with a strategy for the ordinary cube already form a strategy for the supercube. The brief supercube maneuvers assembled in figure 2 permit a particularly economical strategy. The structure of the four 8-move maneuvers m_{802}, m_{803}, m_{805} and m_{806} demonstrates

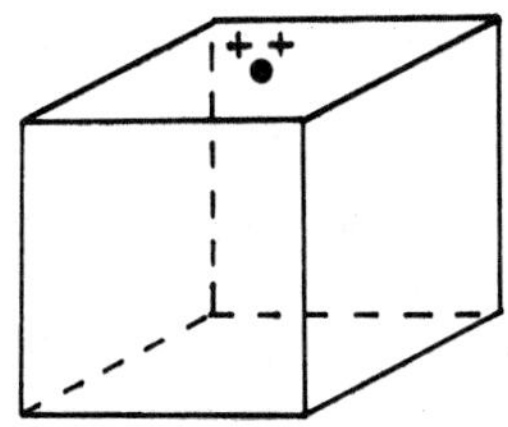

$m_{800} = (URLU^2R'L')^2(12) \rightarrow (++u)$

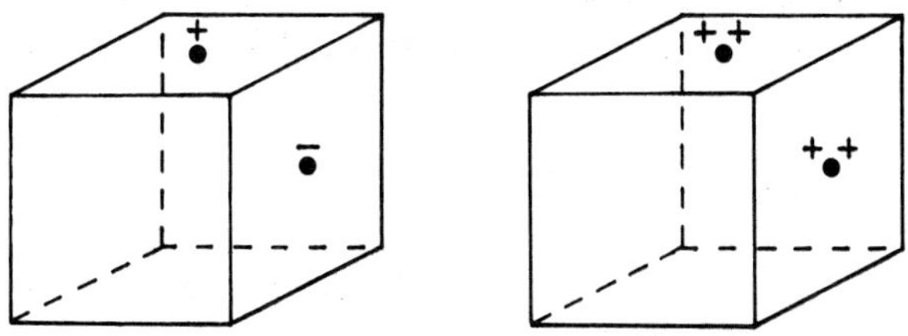

$m_{802} = M'_R M'_D M_R \cdot U'_8 M'_R M_D M_R \cdot U(8) \rightarrow (+u)(-r)$
$m_{803} = M'_R M'_D M_R \cdot U^2 \cdot M'_R M_D M_R \cdot U^2(8) \rightarrow (++u)(++r)$

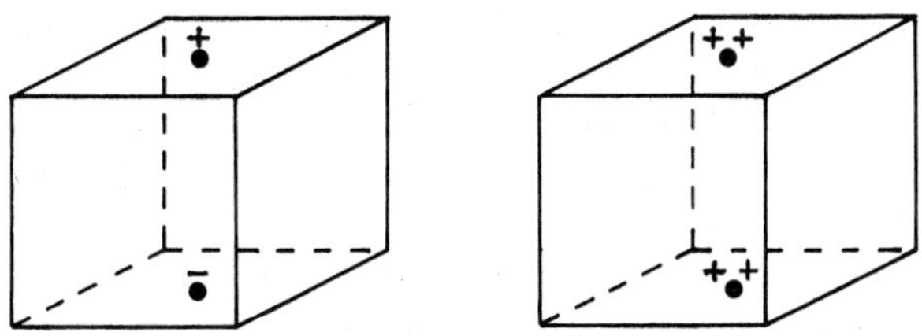

$m_{805} = M'_R M^2_D M_R \cdot U' \cdot M'_R M^2_D M_R \cdot U(8) \rightarrow (+u)(-u)$
$m_{806} = (M'_R M^2_D M_R \cdot U^2)^2(8) \rightarrow (++u)(++u)$

FIGURE 2. Five short Supercube maneuvers

an attractive new application of the isolation principle discussed in section 3.4.

In a mathematical model, the turns of the face cubies can be treated analogously to the characterization of the different "positions" of the corner and edge cubies. For this purpose, we imagine the line markings of the supercube in its start position transferred onto the fictitious, immovable, transparent second skin we already used in section 2.3. At the same time, we label the 4 sides of every face cubie ("sides" here in the sense of the sides of a square) clockwise by 0, 1, 2 and 3 in such a way that the marked side receives the number 0. After choosing an arbitrary numbering for the 6 face cubies, we can then describe their position by a 6-tuple $z = (z_1, \ldots, z_6) \in Z := \{0,1,2,3\}^6$, where z_k tells us, which of the 4 sides of the kth face cubie is lying beneath the line marking of the fictitious skin ($k = 1, \ldots, 6$). Thus, we arrive at the following

Definition. *A position of the supercube* is a 5-tuple (ρ, σ, x, y, z), (ρ, σ, x, y) being a position of the cube and $z \in Z = \{0,1,2,3\}^6$. It is called *possible* when it can be reached from the start position $(1, 1, 0, 0, 0)$ of the supercube by outer layer turns. (Out of consideration for the reader, we do without the obvious mathematical formalization of this statement.)

Theorem 1. *A position of the supercube is possible, precisely if the conditions (a), (b) and (c) of theorem 2.3.3 and the condition*

$$\text{(d)} \quad (-1)^{z_1+\ldots+z_6} = \operatorname{sgn} \rho$$

are fulfilled.

Proof. If (ρ, σ, x, y, z) is a possible position of the supercube, then (ρ, σ, x, y) is a possible position of the ordinary cube, i.e. the conditions (a), (b) and (c) of theorem 2.3.3 are fulfilled. Equation (d) is necessary, because it is fulfilled for the start position $(1, 1, 0, 0, 0)$ of the supercube (here $z_1 = \ldots = z_6 = 0$ and sgn $= \operatorname{sgn} I_{S_8} = 1$) and is preserved in each of the outer layer moves F,B,R,L,U and D: Each of these moves changes the sign of (corner 4-cycle) and one of the summands in the sum $z_1 + \ldots + z_6$ by 1. That (a), (b), (c) and (d) are sufficient follows from Theorem 2.3.3 and the considerations above. □

From this we immediately get the promised

Theorem 2. *There exist precisely*

$$2^{11} \cdot P = 88\ 580\ 102\ 706\ 155\ 225\ 088\ 000 = 2^{38}3^{14}5^{3}7^{2}11$$

(more than 88 sextillion) possible positions of the supercube.

If in Section 2.1 the dimension of a bridge constructed out of all the cubes with different possible positions could still be illustrated by distances as they exist within our galaxy, we must–in the case of the supercube–move into intergalactic space. $88.58 \ldots \cdot 10^{21}$ little supercubes laid side by side make up a distance of over 500 000 light-years and that amounts to appr. one fourth of the way to Andromeda. Tightly packed, the supercubes–every one of them in a different position–would cover the surface of the earth ($51 \cdot 10^{17}$ cm^2) to a height of ca. 30.5 km. An impressive time comparison, useful to judge computer applications, has been contributed by D. Singmaster: If a computer were to count all the supercubes, counting one million per second, it would need appr. one third of the total age of the universe (since the hypothetical big bang).

If one considers the "position" of the cube relevant, we get a concept of the position where the three types of cubies are completely equal: A *position of the supercube* is a 6-tuple $(\rho, \sigma, \tau, x, y, z)$ with $\rho \in S_8$ $\sigma \in S_{12}, \tau \in S_6, x \in X = C_3^8$, $y \in Y = C_2^{12}$, $z \in Z = C_4^6$.

5.2 Sam Loyd's 15-Puzzle

It has happened once before, in the seventies of the nineteenth century, that a puzzle conquered the world and, above all, America. It was invented by Sam Loyd, an amazingly versatile and productive man who invented puzzles of all kinds. The anecdotes connected with that puzzle–still one of the favorite prizes at children's parties–are indeed incredible. Merchants did not open their stores, people did not eat for days, and betting enthusiasts lost fortunes.

Yet, compared to Rubik's Cube it is a rather modest object. Fifteen consecutively numbered tiles can be moved around a 4×4 frame containing always one empty space by shifting one of the tiles next to this empty space into it. Before the game developed the shape that we are used to today, it simply consisted of 15 numbered dice-shaped wooden pieces that were shifted around a box.

Figure 1 shows the "start position": the tiles are numbered in rows from 1 to 15 and the empty space is at the bottom on the right. In every other position with the empty space at the bottom right some or all of the tiles have exchanged places. The position corresponds therefore to a permutation of the numbers 1 to 15, hence to an element of the symmetric group S_{15}. Figure 2 shows the cunningly conceived position in which the game was often sold during Sam Loyd's time. It originates from the start position by swapping 14 and 15, and is, therefore, an odd permutation. That it was not altogether accidental that the puzzle freaks of that time, attempting to find their way back to the start position, were driven to the verge of insanity is shown by the following

Theorem. *The positions with the empty space at the bottom right that can be reached from the start position of Sam Loyd's 15-Puzzle by shifting tiles are precisely the* $15!/2 = 3 \cdot 4 \cdot 5 \cdot \ \ldots \ \cdot 15 = 1\,307\,674\,368\,000$ *even permutations of the numbers from 1 to 15.*

FIGURE 1. Sam Loyd's 15-Puzzle

1	2	3	4
5	6	7	8
9	10	11	12
13	15	14	

FIGURE 2.
An impossible position for Sam Loyd's 15-puzzle

Proof. The empty space is denoted by the number 16. Every time an individual tile is shifted around, the 16 is exchanged with one of the neighboring tiles, and every condition of the puzzle (the empty space can be anywhere) can be seen as an element of the symmetric group S_{16}.

(a) With every sequence of shifts transferring the start position into another position with the empty space at the bottom right, the number 16 follows a closed path. It contains the same number of "north" and "south" steps and as many "west" as "east" steps. The path consists therefore of an even number of individual steps (2 × the number of the north steps and 2 × the number of the west steps), we perform an even number of transpositions, and the permutation of S_{16} which we reach is even. Since the number 16 returns to its original place at the very end (it is a "fixed point" of the permutation) the element of S_{15}, reached by cancelling the number 16, is also even. Summary: *At the most,* all even permutations can be reached.

(b) If the three tiles surrounding the empty space at the bottom right are shifted around in a circle following the order of moves south-east-north-west, the 3-cycle (11, 12, 15) is produced, brief *SENW* (11, 12, 15). Summary: It is easy to reach a *special* 3-cycle.

(c) Assume m to be a "maneuver" bringing three given tiles a, b and c to the places (cubicles for tiles) 11, 12 and 15, while all the other tiles can be changed at will. Analogously to the cube we denote by m' the "inverse maneuver" which annuls m. We then have $m \cdot SENW \cdot m' \rightarrow (a, b, c)$ (conjugation in the group S_{16}). Summary: Every 3-cycle, and therefore every even permutation (theorem 2.2.2) can be reached. □

Theorem and proof are obviously applicable to any rectangular puzzle of the size $r \times s$ with $r \geq 2$ and $s \geq 2$.

Part (a) of the proof calls to mind a well-known chess problem by K. Fabel. Since cubology has a certain number of traits in common with problem chess, we shall—for the benefit of chess experts—briefly describe this problem: White: K-QB8, R-K1, N-Q1, N-K2, P-QN2, QN5, QB6, KN3 (8 pieces). Black: K-KR8, R-KR7, B-QR2, B-KN7, N-QR1, N-KB8, N-KR6, P-QN6, QN3, QB2, KN5, KN3, KN2, KR2 (14 pieces). With White beginning, we have checkmate in 182 moves. Neither the third black knight (exchange piece), nor the number of moves (one hundred and eighty-two) are misprints. The white king must follow some closed paths from d8 back to d8. This requires an even number of moves if he always stays on black squares.

5.3 Other Magic Polyhedrons

It is tempting to transfer the fascinating technique of the magic cube to other solids. Regular polyhedrons seem particularly suited for this purpose. A *polyhedron* is a body with flat boundaries. It is called *convex* if all its corners point outward. A *regular polyhedron or platonic solid* is a convex polyhedron bounded by regular polygons congruent to each other, equally many of which meet at every corner. It is obvious that regular polygons with 6 or more corners are not suitable for building regular polyhedrons: Their internal angle is at least 120° (hexagon), hence three or more of them cannot fit into one corner. The same argument proves that not more than 3 regular pentagons (internal angle 108°), not more than 3 squares (internal angle 90°) and not more than 5 equilateral triangles (internal angle 60°) can meet in one corner. This leaves precisely five cases which can all be realized, and this can be easily checked (figure 1):

(a) 3 equilateral triangles at every corner produce the *tetrahedron* (4 faces, 6 edges, 4 corners).

(b) 4 equilateral triangles at every corner produce the *octahedron* (8 faces, 12 edges, 6 corners).

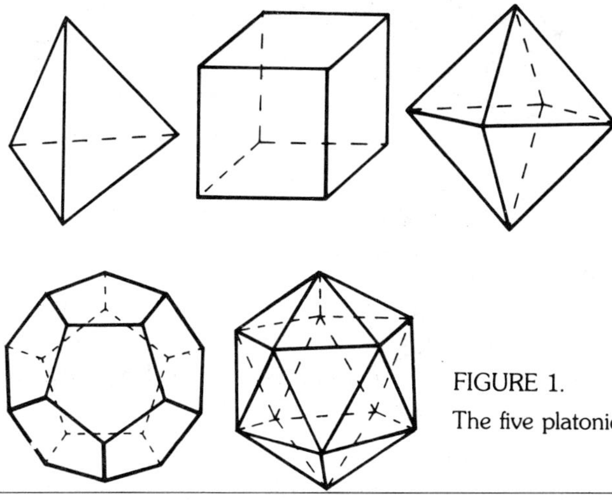

FIGURE 1.
The five platonic solids

(c) 5 equilateral triangles at every corner produce the *icosahedron* (20 faces, 30 edges, 12 corners).

(d) 3 squares in every corner produce the *hexahedron* (6 faces, 12 edges, 8 corners). It has reached a certain amount of notoriety under its everyday name "cube".

(e) 3 regular pentagons at every corner produce the *dodecahedron* (12 faces, 30 edges, 20 corners).

We now cut a platonic solid with as many planes as it has faces, with each of the cutting planes running parallel to one of the faces. The distance between the plane and the corresponding face should be small and the same for all the planes. Thus, the platonic solid is divided into a central piece (a smaller platonic body of the same type), corner pieces congruent to each other, edge pieces congruent to each other, and face pieces congruent to each other. As with the cube, we can ensure that all the pieces stay attached and each of the layers consisting of a face piece and the neighboring edge and corner pieces can be turned around its symmetry axis. This is again done by equipping the individual parts with corresponding "feet" and gaps. This is sometimes tricky,

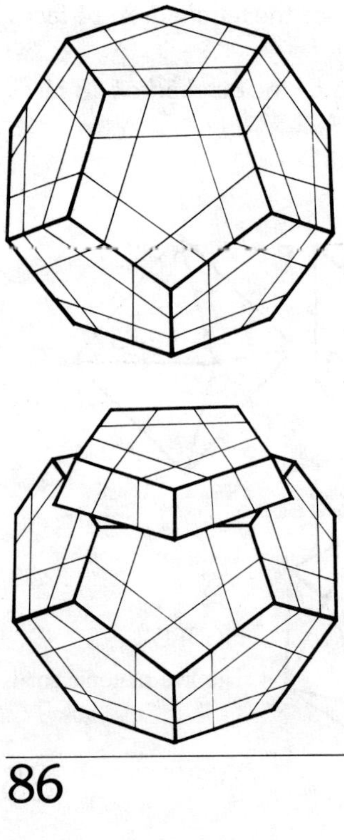

FIGURE 2. The Magic Dodecahedron

but we have reason to believe that some of the *magic platonic solids* differing from the cube are now also available.

As an example, we take a closer look at the *magic dodecahedron* which–due to its harmonic form–is the queen of all magic solids (figure 2). Here 12 different colors can be mixed and then "sorted out", if in the start position each of the faces has one color and the colors of all the faces differ. It is easy to see that by suitable turns of the layers by integral multiples of 72° each corner piece of each of the 20 corners can be shifted to every other corner and each edge piece of every one of the 30 edges can be moved to every other edge. As with the cube, each corner piece can assume 3 "positions" at every corner and each edge piece 2 "positions" at every edge. A possible position is a color pattern which can be reached from the start position by turning the 12 (outer) layers.

Theorem 1. *The magic dodecahedron has precisely*

$$d := 20! \cdot 30! \cdot 3^{20} \cdot 2^{30}/(2 \cdot 2 \cdot 2 \cdot 3) = 20! \cdot 30! \cdot 3^{19} \cdot 2^{27} \approx 10^{68}$$

possible positions.

Proof. Due to the analogous proof of the corresponding cube theorem we can be brief.

(a) Every 72°-turn of a layer produces a 5-cycle for corner pieces and a 5-cycle for edge pieces. Therefore, only even permutations of corner pieces and even permutations of edge pieces are possible. A marking procedure similar to that used for the cube shows that the "positions" of 19 corner pieces and 29 edge pieces determine the "position" of the 20th corner piece and the 30th edge piece. Hence, the given number of positions *at the most* is possible.

(b) By isolation (cf. Section 3.4) which is easier to realize on the "spacious" dodecahedron than on the cube, we can construct a 3-cycle for corner pieces, a 3-cycle for edge pieces, the turning of two corner pieces (opposite direction) and the turning of two edge pieces: Then, conjugation and combination (Sections 3.3 and 3.2) quickly show that at least the given number of positions is possible. □

By adding 12 small lines we can, similar to the cube, transform the magic dodecahedron into a super dodecahedron, where the "position" of the face piece is relevant.

Theorem 2. *The superdodcahedron possesses precisely*

$$5^{12} \cdot d = 5^{12} \cdot 20! \cdot 30! \cdot 3^{19} \cdot 2^{27} \approx 2.46\ 10^{76}$$

possible positions.

Proof. We want to show that a single face piece can be turned by 72° without changing anything else.

The maneuvers for corner and edge 3-cycles constructed in part (b) of the proof of Theorem 1 by means of isolation and conjugation are commutators of the dodecahedron maneuver group (cf. proof of Section 2.5.3). Therefore, for each such move the inverse move is also a commutator; they both leave the face pieces unchanged. This property is shared by corner and edge 5-cycles which result from a combination of two of these 3-cycles (cf. Section 3.2). We can, therefore, turn a face piece by turning the whole layer and afterwards turning the corner and edge pieces back by means of 5-cycles which do not change any face piece. □

Let us take another look at the supercube of Section 5.1. Every corner cubie can assume 3 different "positions" in 8 different spots, all in all, therefore, $8 \cdot 3 = 24$ "positions". Analogously, we have for every edge cubie $12 \cdot 2 = 24$ and for every face cubie $6 \cdot 4 = 24$ "positions", and 24 is also the order of the rotation group D of the cube. Due to the special symmetry properties of these solids, we have a corresponding theorem for each of the 5 "super platonic solids". In the case of the super dodecahedron every corner piece, every edge piece and every face piece can assume 60 different "positions" and this is the order of the rotation group of the dodecahedron which is isomorphic to A_5.

The following examples prove that we do not necessarily have to assign a movable part to every corner, edge and face and that platonic solids are not the only shapes under consideration.

The $2 \times 2 \times 2$ cube (figure 3) is already extensively described in Rubik's patent specifications of 1975. It consists exactly of the corner cubies of the ordinary cube. That part of a strategy for the large cube which is dealing with corners is, therefore, always a strategy for the small cube. To obtain a fixed orientation we assume that one of the corner cubies, as f.i. dbl, always remains fixed. (Every turn of the bottom layer can be replaced by one of the top layer, every turn of the back layer by one of the front layer, and every turn of the left layer by one of the right layer.) The remaining seven corner cubies can now be permuted at will, since every maneuver which produces a corner trans-

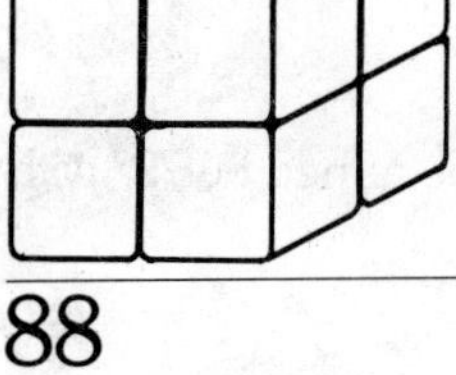

FIGURE 3. The 2 x 2 x 2-Cube

position and an arbitrary edge permutation, will produce nothing but a corner transposition on the small cube. The restriction with regard to the orientation remains unchanged and we get

Theorem 3. *The* $2 \times 2 \times 2$ *cube has precisely* $7! \cdot 3^6 = 3\,674\,160$ *possible color patterns.*

The *Magic Domino* (color plate 2) has been manufactured in Hungary since 1979. Contrary to the $2 \times 2 \times 2$-cube it requires some new strategic thinking, since only those cube maneuvers which do not contain 90° moves with vertical layers can be realized. The most important are

$$m_{620} = (R^2U^2)^3 \ (6) \rightarrow (uf,\ ub),$$
$$m_{100a} = R^2F^2UB^2U'F^2UB^2U'R^2 \ (10) \rightarrow (ufl,\ urf,\ ubr),$$
$$m_{100a}U = R^2F^2UB^2U'F^2UB^2U'R^2U \ (11) \rightarrow (ulb,\ ubr)\ (ub,\ ur,\ uf,\ ul).$$

By means of the last maneuver, we achieve relatively easily every permutation of the 8 corner cubies (conjugation) without paying attention to the 8 edge cubies. By means of m_{620} these can afterwards be rearranged at will. However, since we must not count rearrangements which can also be reached by turning the whole domino around its vertical axis, we get

Theorem 4. *The Magic Domino has precisely* $(8!)^2/4 = 406\,425\,600$ *possible patterns.*

5.4 Mèffert's Pyraminx

An attractive puzzle in the shape of a tetrahedron has been invented by the Swiss, Uwe Mèffert (photograph). It is simpler than the magic tetrahedron described in the last section and considerably simpler than Rubik's Cube. Under the popular name "Pyraminx"–evoking associa-

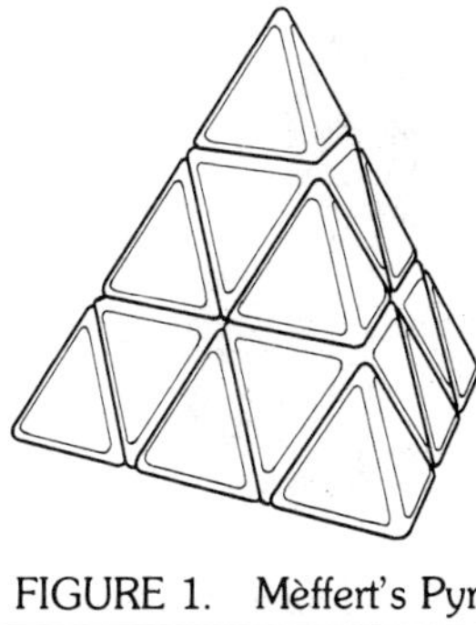

FIGURE 1. Mèffert's Pyraminx

tions of the famous Egyptian pyramids and the inscrutable sphinx–it reaches higher and higher sales records. The fact that the grandiose monuments of the Pharaohs have a square ground-plan seems not to interfere with this success.

Mèffert's Pyraminx (figure 1) consists of the following components:

(a) 1 center ball which is not visible from the outside.

(b) 4 octahedron-shaped "core pieces", attached to the center ball in such a way that they remain turnable, playing the part of the face cubies of Rubik's Cube.

(c) 6 tetrahedron-shaped "edge pieces" kept in place by the octahedrons, playing the part of the edge cubies of Rubik's Cube.

(d) 4 tetrahedron-shaped "peaks", each of which is turnable on one of the core pieces, but otherwise firmly fixed.

This description already shows the essential structural difference between Mèffert's Pyraminx and Rubik's Cube: The cubies of Rubik's Cube are attached by a two-step hierarchy–face cubies hold edge cubies and edge cubies hold corner cubies–while the Pyraminx is held together by a one-step hierarchy: the core pieces hold the edge pieces, while these in their turn have nothing to hold. Every peak can turn independently of everything else on the core piece assigned to it. Similar to Rubik's Cube, an edge piece can be carefully removed from its tracks, if one of the two layers the edge piece belongs to is first rotated by 60°. After all the edge pieces are removed, the skeleton structure shown in figure 2 is left over.

It proves convenient to hold the Pyraminx in such a way that one peak faces upwards, another to the left, the third to the right and the fourth to the back. We then denote by u, l, r and b a 120°-clockwise-turn of the upper, left, right and back peak, respectively (seen from the

FIGURE 2. Pyraminx without edge pieces

outside onto the peak in question). U, L, R and B denote the corresponding turn of a "big peak", i.e. the peak together with its neighbouring layer consisting of a core piece and three edge pieces. A prime added to one of these symbols means, as usual, that the turn in question should be performed counterclockwise. Since the "basis layers" opposite the peaks obviously do not have to be turned, the 4 core pieces and the 4 peaks always remain in their places. All we can change is the place and the orientation of the edge pieces, and the orientation of the core pieces and peaks.

Theorem. The number of possible positions of Mèfferts Pyraminx is $\frac{6!}{2} \cdot 2^5 \cdot 3^8 = 75\,582\,720$.

Proof. Every 120°-turn of a big peak causes an edge-3-cycle, hence an even permutation of the edge pieces. Therefore, at the most 6!/2 placements for the 6 edge pieces are possible. Moreover, every 120°-turn of a big peak causes two disjoint 3-cycles, consequently an even permutation, for the set of the 12 colored tiles on the 6 edge pieces. Therefore, in this set too, we get only even permutations, which means that the reorientation of an individual edge piece as a single transposition, hence an odd permutation, is impossible. On the other hand, the following maneuvers show that every even permutation of edge pieces can be realized and that, in this process, 5 of the 6 edge cubies can be oriented at will $\left(\text{factor } \frac{6!}{2} \cdot 2^5\right)$. From t_0 we get, that the four core pieces can be arbitrarily turned (factor 3^4) and the same applies, rather trivially, also to the 4 peaks (factor 3^4). □

The following list contains a selection of an extensive maneuver-collection for the Pyraminx, prepared by Tim Bandelow. It easily provides us with many good strategies. The notation for operations we apply is completely analogous to the one we use for Rubik's Cube and does not require any further explanations.

(a) core twist

$$t_0 = (RUR'U)^2u' \ (9) \rightarrow (+flr)$$

(b) edge flips

$$t_{10} = RU'R'U \cdot R'LRL' \ (8) \rightarrow (+fl)\ (+fr)$$
$$t_{11} = RUR'U \cdot L'U'LU' \ (8) \rightarrow (+fd)\ (+lr)$$
$$t_{14} = (RBRU'RL)^3 \ (18) \rightarrow (+fl)\ (+fr)\ (+fd)\ (+lr)\ (+rd)\ (+dl)$$

(c) edge-3-cycles

$$t_{20} = R'LRL' \ (4) \rightarrow (fl, fr, uf)$$
$$t_{21} = LR'L'R'U'R'U \ (7) \rightarrow (fl, fr, fu)$$

$$t_{30} = RU'R'U'RU'R' \ (7) \rightarrow (fl, lr, rf)$$
$$t_{31} = R'L'U'LUR \ (6) \rightarrow (fl, lr, fr)$$
$$t_{40} = U'R'UR \ (4) \rightarrow (fl, rf, dr)$$
$$t_{41} = UBUB'U \ (5) \rightarrow (fl, rf, rd)$$
$$t_{42} = U'B'RBR'U \ (6) \rightarrow (fl, fr, dr)$$
$$t_{43} = U'RB'R'B'U'B'U' \ (8) \rightarrow (fl, fr, rd)$$

Some interesting pretty patterns are

$t_1 = B'R'BR'(UR)^2U'LU'L'$ (12) (**Four Flowers**),
$t_2 = (UR)^2U'LU'L'B'R'BR'$ (12) (**Four Gay Flowers**),
$t_3 = UB'U'L'B'RBU$ (8) (**Four Christmas-Trees,** R. aus dem Spring),
$t_4 = UB'RLRL'RU'L'U'R'$ (11) (**Tornado 1** or **Superfliptwist**),
$t_5 = T_4rlub$ (15) (**Four Carriage-Wheels 1**),
$t_6 = R'U'B'U'R'L'RU$ (8) (**Tornado 2,** R. aus dem Spring),
$t_7 = T_6rlub$ (12) (**Four Carriage-Wheels 2**).

6 COMPUTER CUBOLOGY

Electronic computers can be used in many ways to help solve or understand Rubik's cube.

It immediately suggests itself to delegate the work of finding good maneuvers to a computer. The computer systematically simulates all 2-move, 3-move, 4-move etc. maneuvers with moves out of a given subset of the set of all moves. As soon as a maneuver fulfills a given criterion, it is printed out, preferably along with the cyclic decomposition of its operation. If we consider, that for each variation class at most one maneuver needs testing, the work diminishes considerably. Thus, we need not examine more than $3^9 + 2 \cdot 3^7 = 11 \cdot 3^7 = 24057$ ten-move maneuvers involving two perpendicular outer layers, which is no problem for a big computer with a suitable program. (Without loss of generality, we alternately turn the right and the upper layer, we start with R and continue arbitrarily (3^9), or we start with R^2, continue with U or U^2 and end with U^2 ($2 \cdot 3^7$). A more careful analysis reduces the number again by almost its half.) Sensible criteria are, among others, not overstepping certain bounds for the number of cubies to be moved, preserving a part of the cube (f.i. a layer, the corners, or the edges), or causing an operation with a given cycle structure. With this last criterion, J. B. Butler seems to have found his 5-move maneuver, $RU^2D'BD'$, for an operation with the maximal order, 1260. A very simple "maneuver program" for the pocket-calculator TI 59 is described by Michel Dauphin; and Morwen B. Thistlethwaite developed a broadly conceived program.

With the help of the computer, mathematicians can tackle special problems of the structure of Rubik's group G, as f.i. determining systems of generators and defining relations.

In this section, we would like to use the computer to solve the game of solitaire itself, that is to bring a mixed-up cube back to the start position. Such a program can be regarded as a demonstration of a solution

strategy, as a popular model program, but, above all, as simply a charming exercise in programming non-numerical problems.

Can we program a computer to find an optimal, i.e. shortest solution maneuver for any scrambled cube by systematically trying out all maneuvers of growing length? This idea is readily seen as completely off the track: Even if a "maneuver-generator" were to work without any redundancy, i.e. exclusively generating pairwise non-equivalent maneuvers, there would be as many of them as there are positions, hence more than 43 quintillions, and therefore even the fastest computer would require more than the brief time granted to a human being on earth.

The situation changes immediately, if the systematic search is restricted to a fixed previously determined *step-ladder*. A step-ladder is a sequence $P_0, P_1, \ldots, P_n$ of sets of positions (*steps*), where P_0 is the set of all possible positions and P_n contains only the start position we aim for. Often, but not always, the P_i (more precisely the sets of operations corresponding to the P_i) are subgroups of G with $P_{i-1} \supset P_i$. We define the *height* of P_i to be the smallest natural number z_i such that for every $p \in P_{i-1}$ there exists an at most z_i-move maneuver m with $pm \in P_i$. If we succeed in finding a step-ladder with sufficiently small heights, where, moreover, it is easy to check whether a given position belongs to a given step, then the computer can find an optimal solution maneuver within the framework of the step-ladder with a maximum of $z_1 + z_2 + \ldots + z_n$ moves by systematic trial. However, developing an appropriate step-ladder and estimating the heights is not essentially different from developing a good strategy and constructing special maneuvers that was already extensively discussed. At this point, we will not pursue this thought further.

Instead, we will describe a program in which the computer follows our well-known "simple strategy" through all the stages. Because of the variety of existing and new or changing program languages, we choose the form of a flow chart, which is clearly and easily understood by all programmers. On the whole, we stick to the generally accepted norm for flow charts. Merely in the use of the stylized rhombus (normally reserved for program-modifications) for branches, and in the design of loops we differ, for reasons of space, from this norm. If, in the case of a branch within a loop, none of the conditions assigned to the possible exits is fulfilled, all the work for the actual value of the loop variable is already done and there is nothing further to do.

The **input** of a position is given as the colors of the 6 · 9 = 54 squares of the cube tiles, which correspond in the program to the 54 alphanumeric variables N1, N2, . . . , N9, F1, F2, . . . , F9, L1, L2, . . . , L9, B1, . . . , B9, R1, . . . , R9, S1, . . . , S9 as shown in figure 1. After choosing arbitrary symbols for the 6 cube colors and an initial position for the cube, the colors of the 9 squares on the upper side of the cube are fed to the computer in rows. The cube moves RIC (Rubikian, cf. section RINIRENEFENEFI), FIC, FIC, FIC, FICRIC applied one after another bring first the front, then the left, then the back, the right and finally the bottom faces to the top, where they are fed in

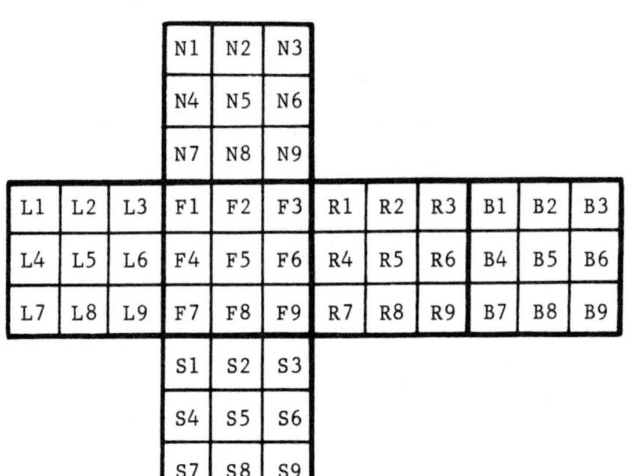

FIGURE 1. The 54 input variables

as the top face before. At the end RUC brings the cube back into its initial position. Schematically we have:

N (RIC) F (FIC) L (FIC) B (FIC) R (FICRIC) S (RUC).

We have written the cube moves required for the input in Rubikian, because this language is very appropriate for the **output,** which we will now describe. Our program first goes through a thorough test of the input. If it does not make sense (f.i. if two face cubies have the same color), the text INPUT ERROR is written. If we are dealing with a genuine magic cube which is merely incorrectly reassembled in such a way that the start position cannot be reached by layer moves, the text CUBE WRONGLY ASSEMBLED will appear, followed by complete repair instructions, consisting of a suitable selection from the following lines:

SWAP ANY TWO CORNER CUBIES OR ANY TWO EDGE CUBIES!
TWIST ANY CORNER CUBIE CLOCKWISE!
TWIST ANY CORNER CUBIE COUNTERCLOCKWISE!
FLIP ANY EDGE CUBIE!

Swapping two cubies always means performing an orientation preserving 2-cycle, i.e. the individual faces of the cubies must exchange their places individually, as f.i. in figure 2. Two edge cubies can be exchanged in 2 and two corner cubies in 3 ways, while preserving their orientation. A reorienting edge-2-cycle (fig. 3) executes a swap of two edge cubies together with a flip of an edge cubie, i.e. the first and the fourth of the above instructions are carried out simultaneously. For corner cubies the situation is analogous.

The following pointers may prove helpful when translating the flow chart into a program:

1. All subroutines are considered as non-parametrized program sections with direct access to the data material of the main program. With the exception of the two subroutines SEC and SEE ("search corner", "search edge") appearing in the test part of the program, the name of every subroutine is the Rubikian symbol of a move, to which

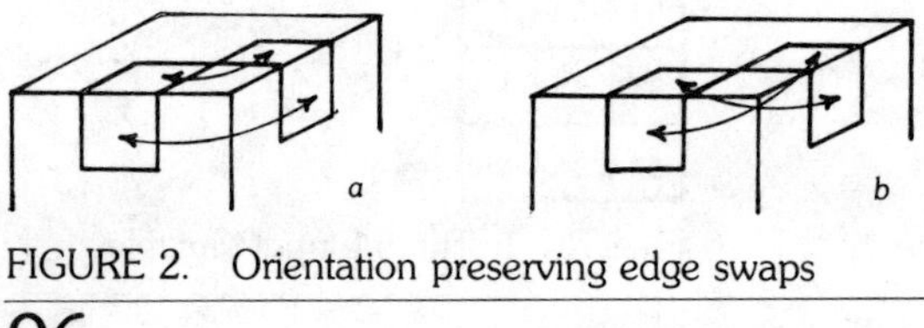

FIGURE 2. Orientation preserving edge swaps

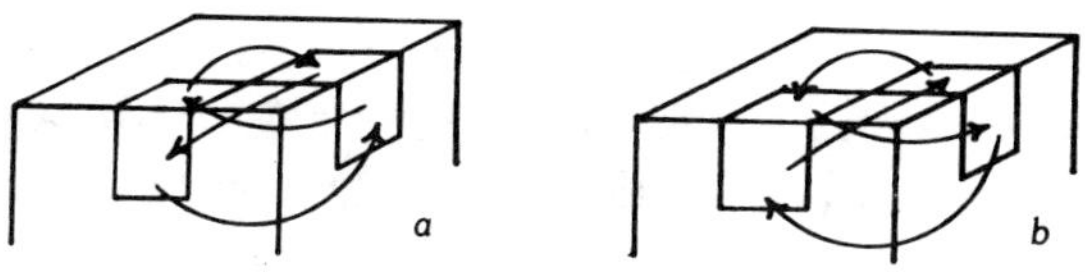

FIGURE 3. Reorienting edge swaps

we occasionally add the letter S. The subroutine NI performs the switching of storage required for the move NI (five 4-cycles) and prints the syllable NI. The subroutine NIS performs the switching of storage without the output of the syllable NI ("simulation of the move NI"). All the other move subroutines are handled analogously; there is therefore no need to write them down in detail. If computer time is no consideration, all the moves can be assembled from a few elements, f.i. from the five components NIS, SEMS, SES, RIS, RIMS: NICS ~ NIS-SEMS-SES, RES ~ RIS-RIS-RIS, LES ~ NICS-NICS-RES-NICS-NICS, RICS ~ RIS-RIMS-LES etc.

2. In many situations it is considerably more economical, to realize the change of position resulting from a sequence of moves directly and with little switching of storage, and to give the output of the corresponding maneuver, than to call up a long sequence of move subroutines. We have, however, decided against this, because that way the flow chart is more easy to understand. Furthermore in this form it can be read as a precise and complete description of an algorithm without any intention of programming.

3. If the realization on smaller computers runs into capacity problems one can omit the test part of the program going from * to ** without further change. Moreover, the part from ** to *** can be scratched without compensation, if the bottom layer of the cube is already completed. In this reduced form, R. Gall even managed to realize the program on a TI 59 (average computing time 3 minutes and 10 seconds) using a few tricks (output of numbers for moves and entire maneuvers, NIM instead of NIC, etc.). The complete program was written in FORTRAN IV by F. Inkmann and needs an average of 0.3 seconds per cube on the old TR 440 of the Ruhr-Universität Bochum, which is being replaced by a considerable faster system. To test the comprehensibility of the flow chart, R. Bürger, pupil of the Albert-Einstein-Schule in Bochum, realized the program in BASIC, on purpose without any knowledge of the strategy. His Alphatronic P2 needs an average of 3 seconds.

4. Without the test part (* to **), apart from the 54 alphanumeric

variables N1, N2, . . . one needs only the two alphanumeric variables A and B, the alphanumeric 4 × 2-array C and the counting variables I and Z. In addition the test part uses the alphanumeric variable D, the alphanumeric 8 × 3-array C (instead of the 4 × 2-array), the alphanumeric 12 × 2-array E, the counting variables J, I1, I2, I3, I4, I5, ZC, ZE, IC, IE and the 1 × 8 or 1 × 12 counting arrays PC ("permutation of the corners") or PE ("permutation of the edges").

Ready-made realizations of this program are available for several computer types. (For details contact the author.)

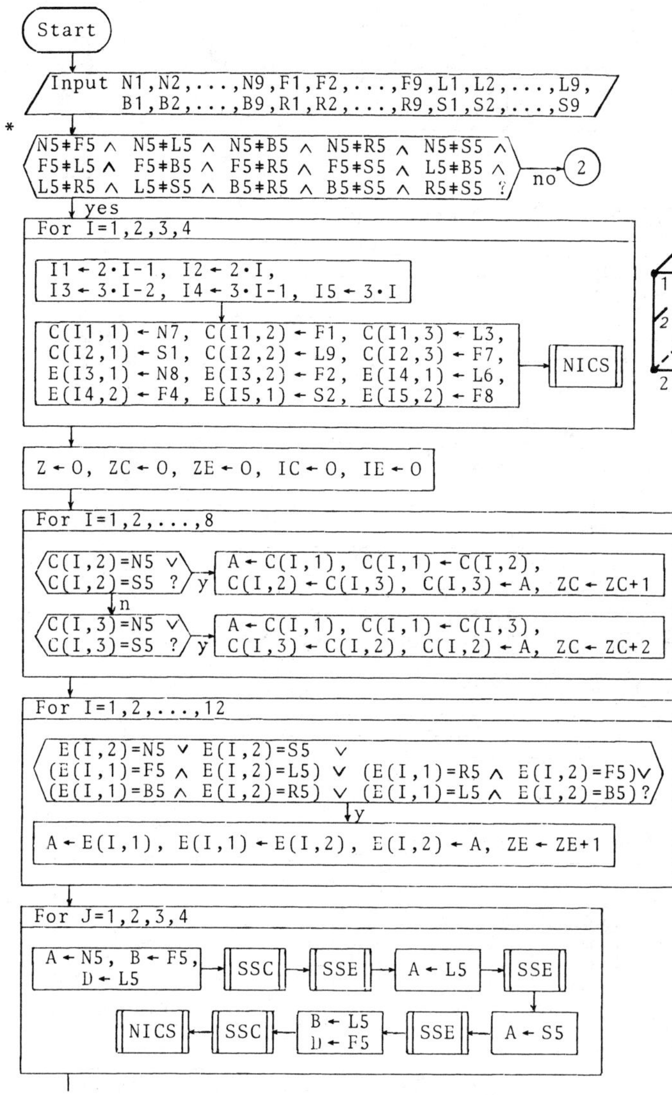

Input and control of the diversity of all the face cubies

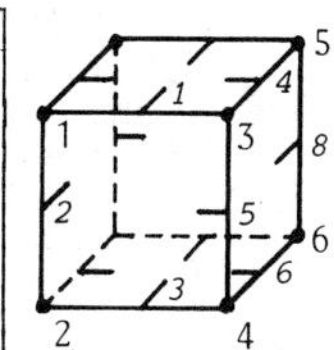

Corner and edge cubies will be listed for the search subroutines *SSC* and *SSE* in standard form (8x3-array *C* and 12x2-array *E*, marked side first; see sketch for the order and edge markings).

Control of the existence of all the corner and edge cubies

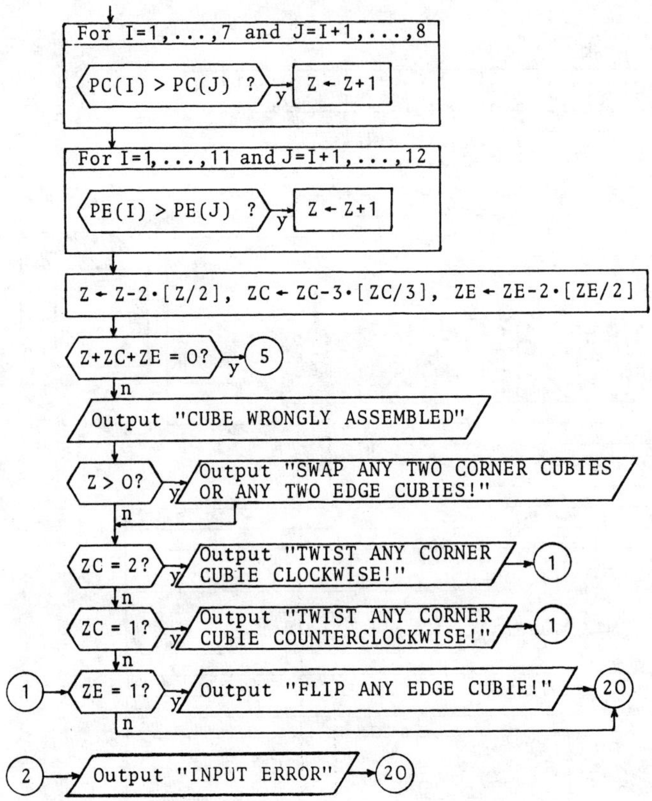

The corner and edge cubies will be searched for in the order in which they would appear in the lists C and E, respectively, if the cube were found in the start position. In this way, we can simultaneously determine the number of inversions, which must be even. **[x]** *is the greatest integer* **≤** *x.*

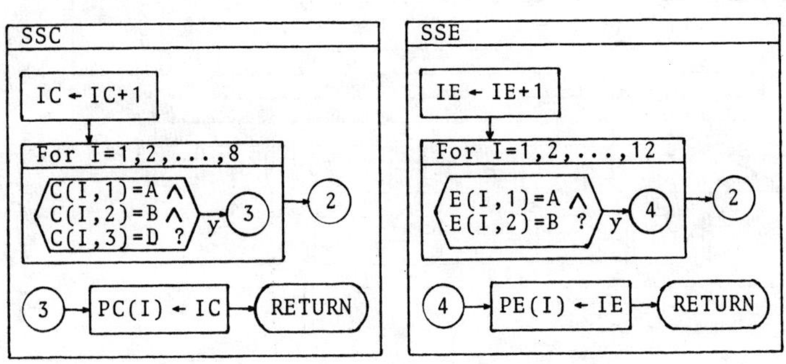

**

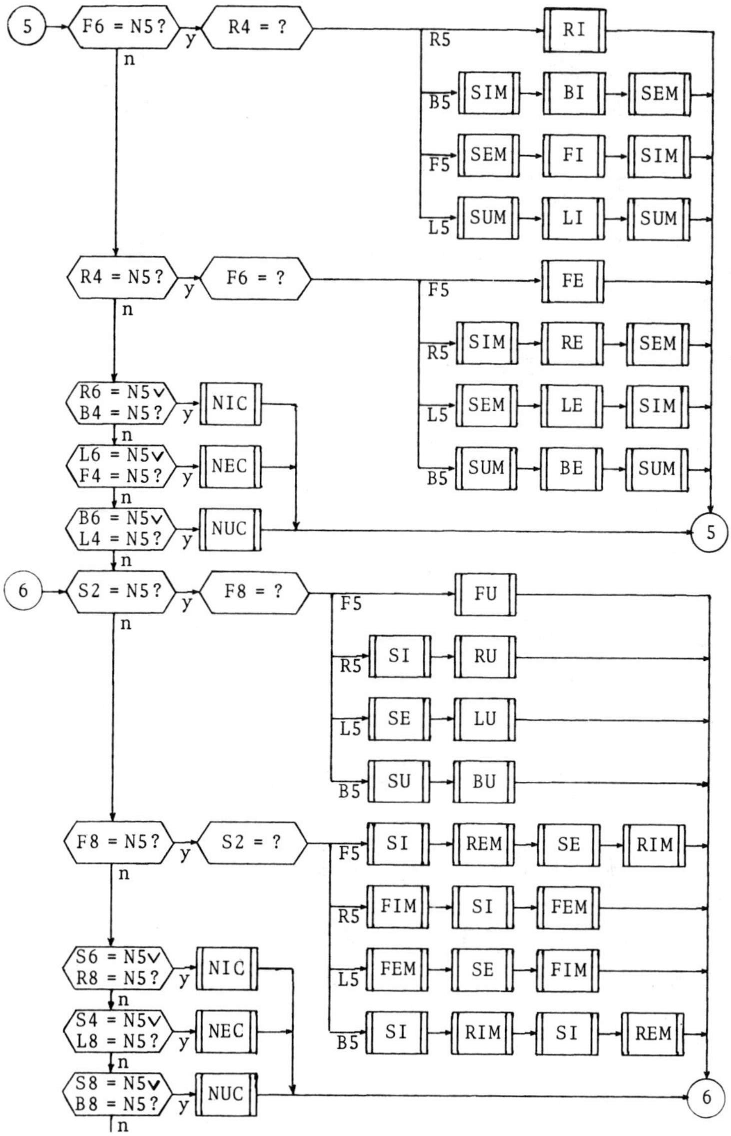

The horizontal middle layer is looked through to find edge cubies which belong to the north layer. They are always first brought to the right-front position and from there to the correct positions.

Here the north edge cubies which were lost in the south, are shown the way home. They always go by way of front-down.

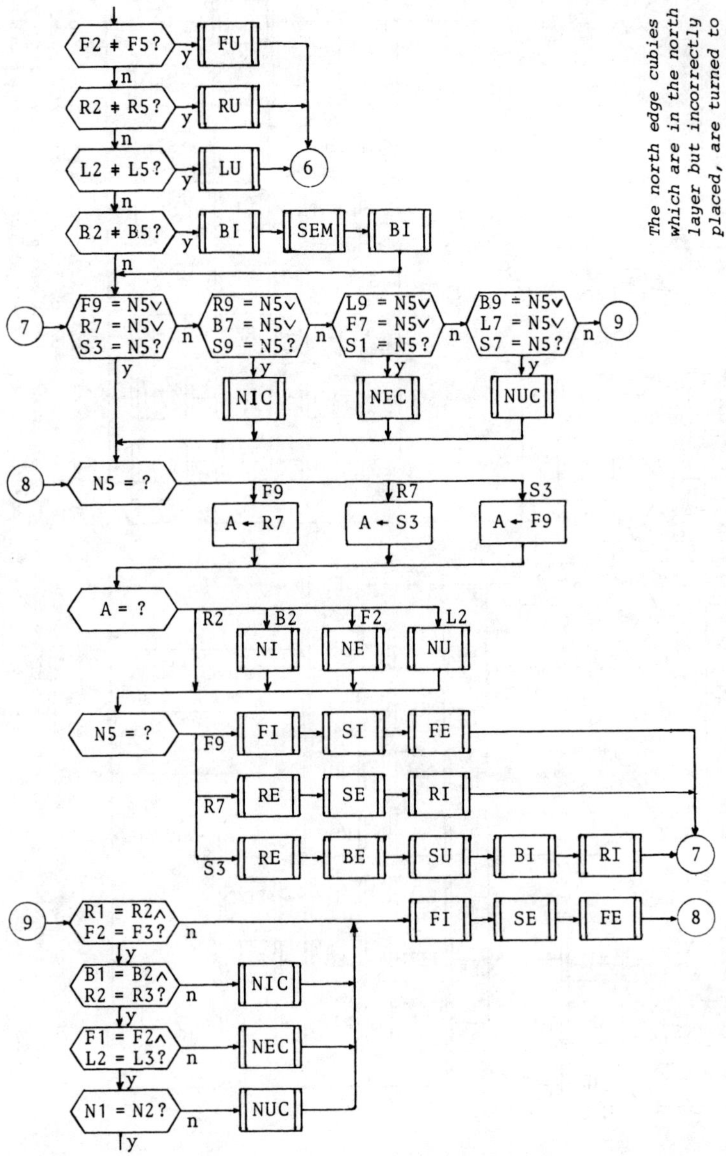

The north edge cubies which are in the north layer but incorrectly placed, are turned to the south, where they encounter the prepared signposts.

Here the corner cubies of the first layer are adjusted. Every misplaced cubie is first brought to right-front-down and its home cubicle to right-front-up. Now the first layer is done.

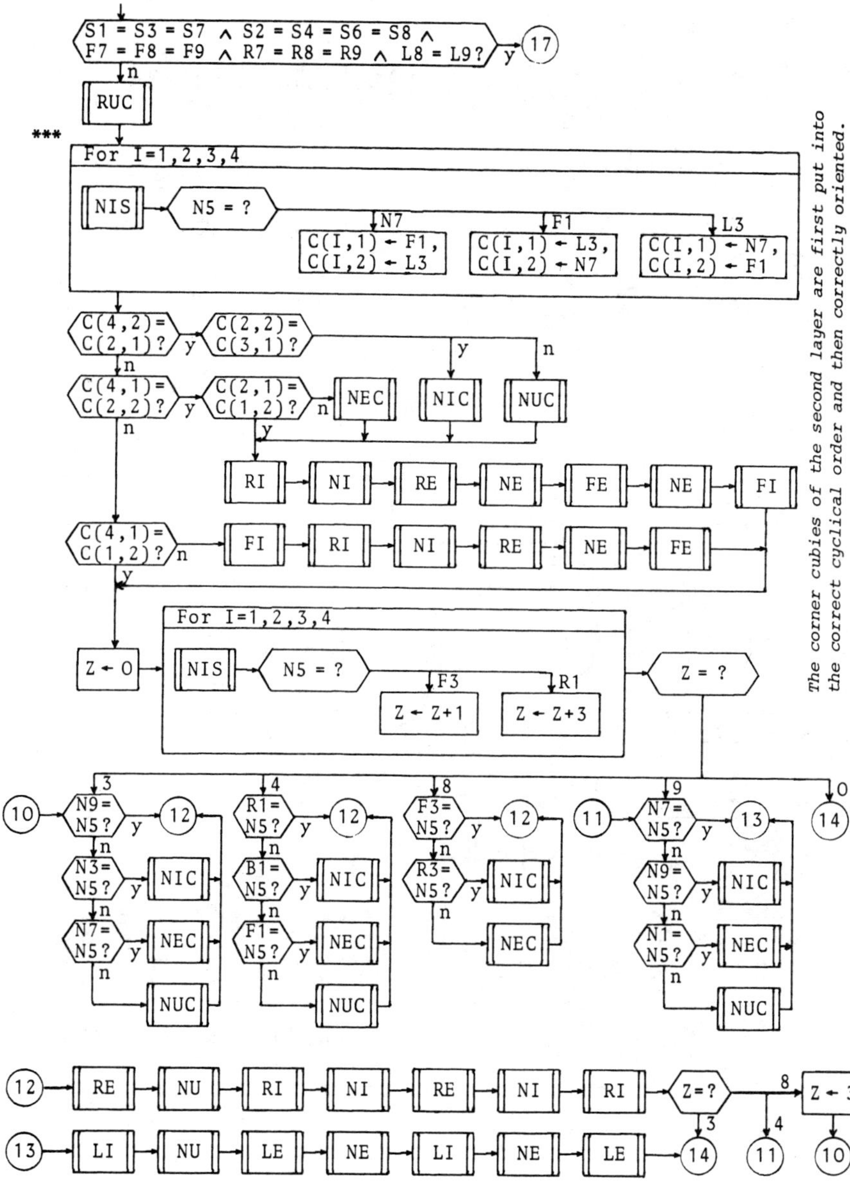

The corner cubies of the second layer are first put into the correct cyclical order and then correctly oriented.

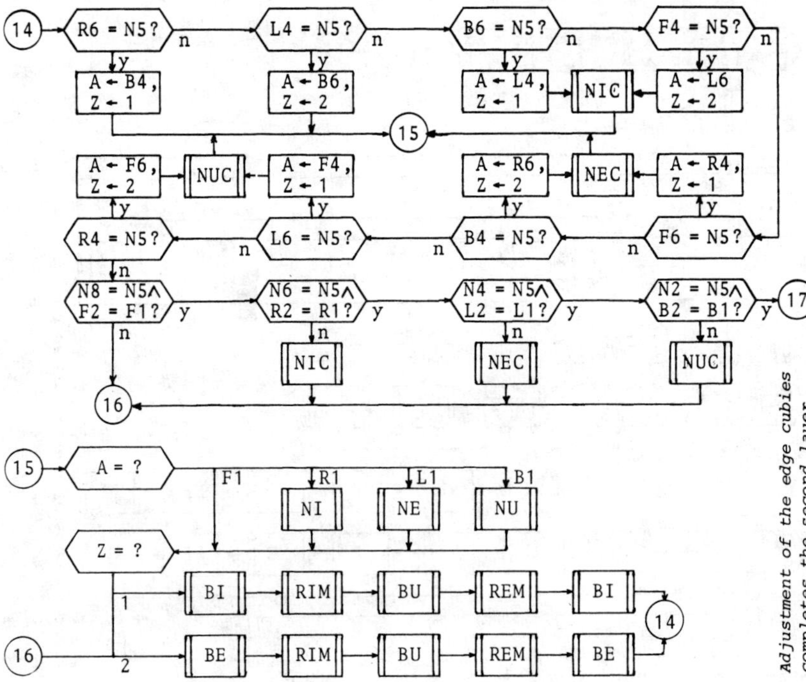

Adjustment of the edge cubies completes the second layer.

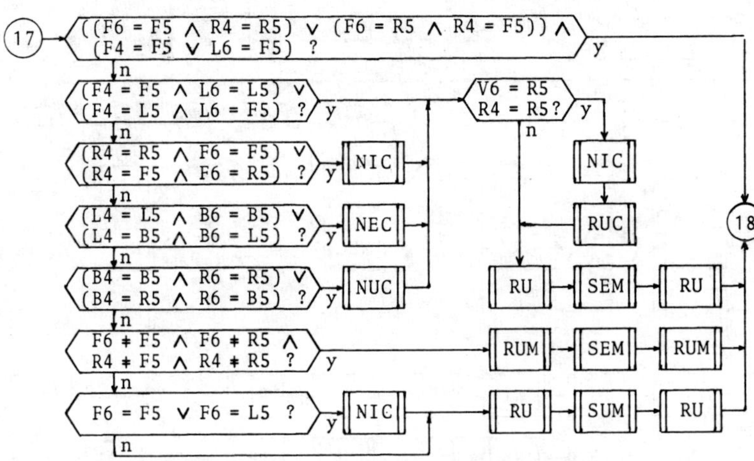

A 3-move-maneuver brings the edge cubies of the middle layer to their places.

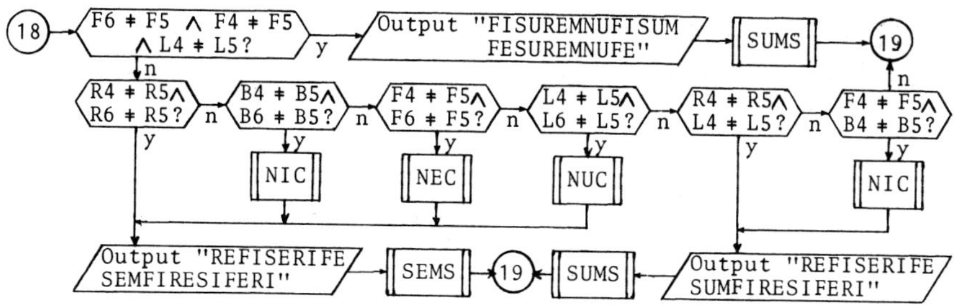

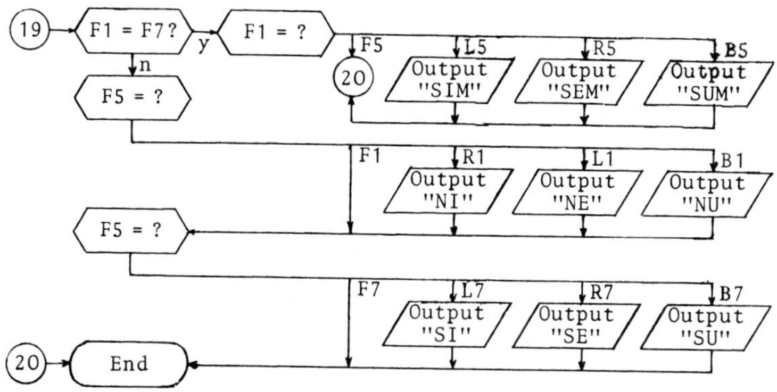

After orienting the last edge cubies, the three layers are lined up.

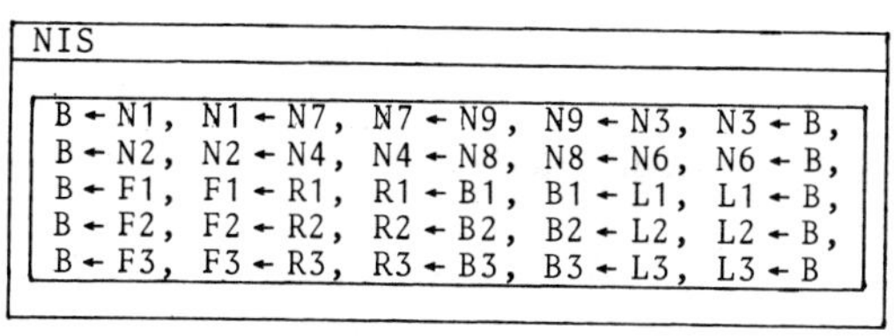

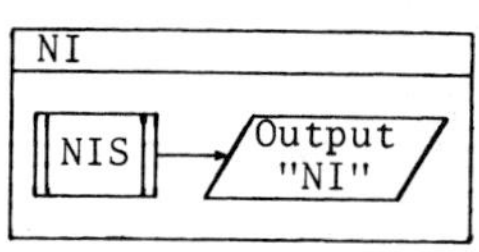

etc.

NIS simulates the move NI without output. NI gives in addition the appropr. output.

MANEUVER INDEX

There exist over 43 quintillion possible positions, just as many possible operations, and for every possible operation we have infinitely many maneuvers. Every selection is, therefore, necessarily incomplete.

However, the following list contains–among others–a complete catalog of 3-cycles; more precisely: for every edge or corner 3-cycle a maneuver is given which causes a variation of the 3-cycle. Since it is impossible to swap two cubies without simultaneously changing the cube somewhere else, 3-cycles play a basic role. More generally, all those operations are represented which do not move more than 3 cubies. The obvious idea to list at the most one maneuver for every variation class was rejected in order to demonstrate all the possibilities for some of the situations and, thus, to obtain the most practical index possible. The maneuvers are classified according to the cycle structure of the operation they cause.

To give fair credit to the authors of individual maneuvers must obviously remain a hopeless enterprise, since most of the beautiful maneuvers have been and will be discovered independently of each other over and over again. Names are therefore mentioned only in a few cases. Special thanks are due to Peter Klerings (Bonn). He contributed so many excellent maneuvers that we simply marked his creations by PK.

A very comprehensive illustrated maneuver collection is contained in the following book: Bandelow, C. and Klering, P.: International Handbook of Rubik's Cube. 1982

(a) Corner twists

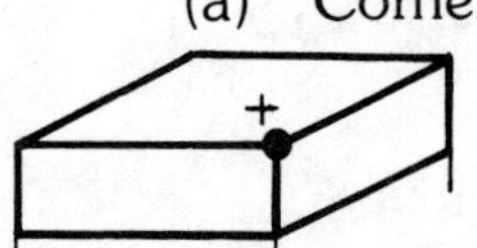

$m_0 - m_4$

m_0 = *R'DRFDF'* (6) → (+*urf*) · (+*dfr*) (+*drb, dbl, dlf*) (*fr, dl, df, rd, bd*)

m_{0a} = $R'DR^2F'R'F$ (6) → (+*urf*) · (−*dfr, ldb, drb, lfd*) (+*fr, db, dl, rd*) (+*fd*) (PK)
(isotwists for the upper layer, cf. Section 3.4)

m_1 = $(RF'R'F)^2$ (8) → (+*urf*) (+*dfr*) (−*dlf*) (−*drb*) (*fr, dr, fd*)
(isotwist of order 3 for the upper layer, cf. Section 3.4)

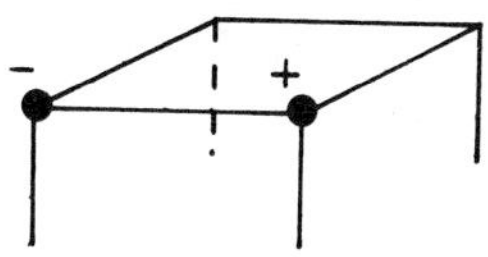

$m_5 - m_{14}$

m_5 = $R(U^2RF'D^2FR')^2R'$ (13) → (+*urf*) (−*ufl*)

m_{5a} = $m_0U'm_0'U$ = $R'DRFDF' \cdot U' \cdot FD'F'R'D'R \cdot U$ (14)

m_{5b} = $C_U' \cdot R'U^2RUR'UR \cdot LU^2L'U'LU'L' \cdot C_U$ (14)

m_{5c} = $UR \cdot U^2RU^2R' \cdot U'RU'R' \cdot U^2R'U^2R \cdot UR'$ (16) (PK)

m_{10} = $F^2R^2B'F'RFR'BR^2F^2U'R'UR$ (14) → (+*urf*) (−*ufl*) · (+*uf*) (+*ur*) (M. B. Thistlethwaite)

m_{11} = $UF^2L^2D'L'DL'FU'F$ (10) → (+*urf*) (−*ufl*) · (*uf, ru, fr*) (PK)

m_{12} = $(F'U)^3(FU')^3$ (12) → (+*urf*)(−*ufl*) · (*fu, ru, bu, lu, fl, fd, fr*)

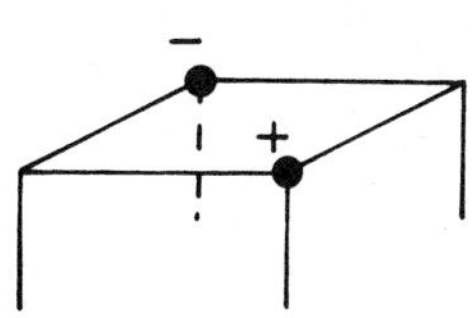

$m_{15} - m_{19}$

m_{15} = $(U^2L'F \cdot C_{LF})^4$ (12) → (+*urf*) (−*ulb*) (E. Rubik)

m_{17} = $R'UF'UFU^2R^2B'R'B$ (10) → (+*urf*) (−*ulb*) · (*ur, ub, rb*) (PK)

m_{20} = $(R'D^2R \cdot C_{FL})^4$ (12) → (+*urf*) (−*dbl*)

m_{22} = $B^2U'B'UB' \cdot RD'RDR^2$ (10) → (+*urf*) (−*dbl*) · (*ur, br, bd*) (PK)

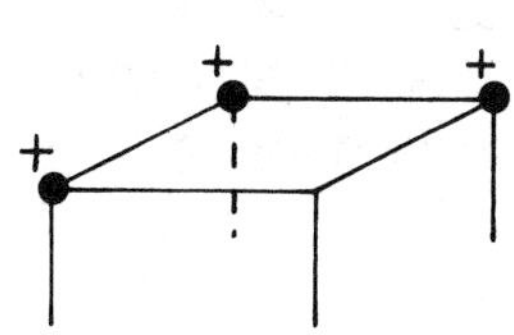

$m_{25} - m_{34}$

m_{25} = $U^2(FR'F^2R)^2U^2 \cdot LFL'F'$ (14) → (+*ufl*) (+*ulb*) (+*ubr*) (M. B. Thistlethwaite)

m_{25a} = $R^2URU'(RU^2RU')^2RUR^2 \cdot U$ (16) (PK)

m_{30} = $R'U^2RUR'URU^2$ (8) → (+*ufl*) (+*ulb*) (+*ubr*) · (*uf, ul, ub*)

m_{31} = $FRB'RBR^2F'U^2$ (8) → (+*ufl*) (+*ulb*) (+*ubr*) · (*ul, ur, bu*) (3-D Jackson)

m_{32} = $(U^2R'U' \cdot C_{UBR})^3$ (9) → (+*ufl*) (+*ulb*) (+*ubr*) · (+*ul*) (+*ub*) (D. J. Benson)

m_{35} = $FU'F'R^2(FD'F^2D)^2R^2U$ (14) → (+*ufl*) (+*urf*) (+*drb*)

m_{40} = $DB'D'F^2RU'R'D'RUR'D^2BD'F^2$ (15) → (+*ufl*) (+*ubr*) (+*dfr*)

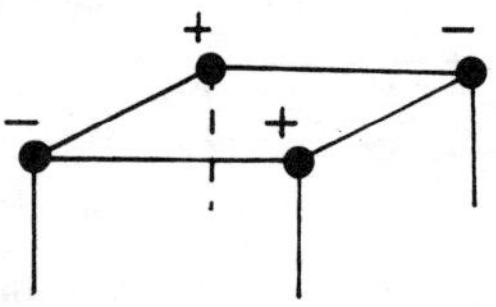

$m_{45} - m_{59}$

m_{45} = $FRB'RBR'B'RBRFM_R^2F'RFM_R^2F^2$ (17) → (+*urf*) (−*ufl*) (+*ulb*) (−*ubr*) (PK)

m_{45a} = $F(URU'BU'B'UR')^2F'$ (18)

m_{45b} = $RU'(RU)^2(R'U')^2R^2U'RUR'U'RU'R'$ (19) (PK)

m_{50} = $R'U^2RUR'U'RUR'UR$ (11) → (+*urf*) (−*ufl*) (+*ulb*) (−*ubr*) · (*uf, ul, ur*)

m_{51} = $R^2U^2RU^2R^2U$ (6) → (+*urf*) (−*ufl*) (+*ulb*) (−*ubr*) · (*uf, ul, fr*) (*ub, ur, br*)

m_{51a} = $L^2D^2RD^2L^2U$ (6) (PK)

m_{52} = $(RUR^2FRF^2UF)^2$ (16) → (+*urf*) (−*ufl*) (+*ulb*) (−*ubr*) · (+*uf*) (+*ul*) (+*ub*) (+*ur*) (D. J. Benson, J. Conway, D. J. Seal)

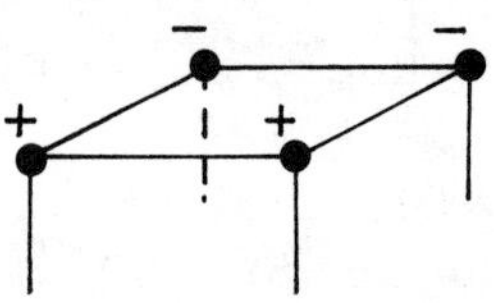

$m_{60} - m_{69}$

m_{60} = $L'BL^2B'LM_F'L^2M_FLU^2B'U^2BL' \cdot U^2$ (15) → (+*urf*) (+*ufl*) (−*ulb*) (−*ubr*) (PK)

m_{60a} = $RU(R^2U')^2R^2U^2R^2U'R'URU^2R'U$ (16) (PK)

m_{65} = $RU^2(R^2U')^2R^2U^2RU^2$ (10) → (+*urf*) (+*ufl*) (−*ulb*) (−*ubr*) · (*ul, ub, ur*)

m_{66} = $U^2R'FR^2B'R^2F'R^2BR'$ (10) → (+*urf*) (+*ufl*) (−*ulb*) (−*ubr*) · (*ur, ul, bu*)

m_{67} = $F(RUR'U')^2F'$ (10) → (+*urf*) (+*ufl*) (−*ulb*) (−*ubr*) · (*uf, bu, ru*)

m_{68} = $(F'U'FRBUB'R')^2$ (16) → (+*urf*) (+*ufl*) (−*ulb*) (−*ubr*) · (+*uf*) (+*ul*) (+*ub*) (+*ur*) (D. J. Benson, J. Conway, D. J. Seal)

m_{69} = $RU'(R^2U)^2R^2U'R \cdot U^2$ (10) → (+*urf*) (+*ufl*) (−*ulb*) (−*ubr*) · (*ul, ur*) (*uf, br, ub, fr*) (R. Brinkmann, P. Dörsam, P. Kellermann)

$m_{75} = RL \cdot U \cdot R'L' \cdot F'B' \cdot U \cdot FB$ (10) → (+*urf*) (+*ulb*) (−*dfr*) (−*dbl*) (*fu, ur, rf*) (*bu, ul, lb*) (R. Walker)

$m_{90} = (U'M_R'FM_R)^3$ (12) → (+*urf*) (+*ubr*) (+*ulb*) (−*ufl*) (−*dlf*) (−*dfr*)

$m_{95} = (U^2L^2R'D)^4$ (16) → (+*urf*) (+*ufl*) (−*ulb*) (−*ubr*) (+*dfr*) (−*dlf*) (+*dbl*) (−*drb*) (M. B. Thistlethwaite)

(b) Corner-3-cycles

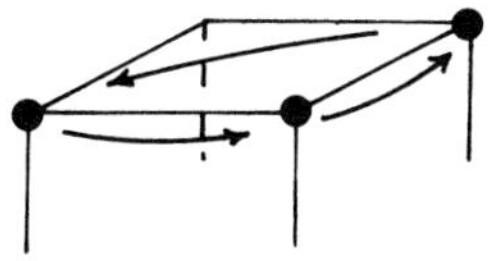

$m_{100} - m_{108}$

$m_{100} = RB'RF^2R'BRF^2R^2$ (9) → (*ufl, urf, ubr*)

$m_{100a} = R^2F^2UB^2U'F^2UB^2U'R^2$ (10)

$m_{101} = R \cdot BL'B' \cdot R' \cdot BLB'$ (8) → (*ufl, urf, bru*)

$m_{102} = F^2D'FU^2F'DFU^2F$ (9) → (*ufl, urf, rub*)

$m_{103} = B'R^2B'L^2BR^2B'L^2B^2$ (9) → (*ufl, rfu, ubr*)

$m_{104} = F \cdot RBR' \cdot F' \cdot RB'R'$ (8) → (*ufl, rfu, bru*)

$m_{105} = F \cdot U'B'U \cdot F' \cdot U'BU$ (8) → (*ufl, rfu, rub*)

$m_{106} = L^2B^2L'F^2LB^2L'F^2L'$ (9) → (*ufl, fur, ubr*)

$m_{107} = L'BU^2B'LBL'U^2LB'$ (10) → (*ufl, fur, bur*) (PK)

$m_{108} = RU^2RDR'U^2RD'R^2$ (9) → (*ufl, fur, rub*)

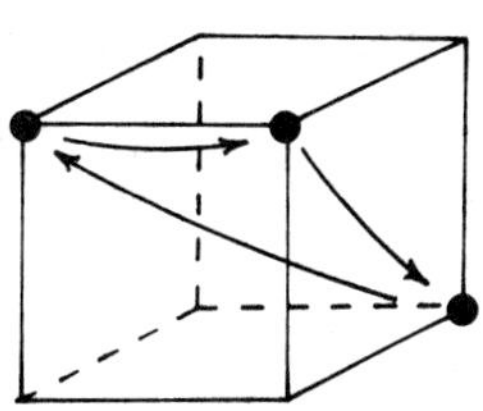

$m_{110} - m_{118}$

$m_{110} = (R^2DL^2D')^2$ (8) → (*ufl, urf, drb*)

$m_{111} = F'D^2F \cdot U \cdot F'D^2F \cdot U'$ (8) → (*ufl, urf, rbd*)

$m_{112} = U' \cdot FD'F' \cdot U \cdot FDF'$ (8) → (*ufl, urf, bdr*)

$m_{113} = UB^2U' \cdot F' \cdot UB^2U' \cdot F$ (8) → (*ufl, rfu, drb*)

$m_{114} = ULU' \cdot R^2 \cdot UL'U' \cdot R^2$ (8) → (*ufl, rfu, rbd*)

$m_{115} = F \cdot U'BU \cdot F' \cdot U'B'U$ (8) → (*ufl, rfu, bdr*)

$m_{116} = L^2DLU^2L'D'LU^2L$ (9) → (*ufl, fur, drb*) (PK)

$m_{117} = L^2B'L'F^2LBL'F^2L'$ (9) → (*ufl, fur, rbd*)

$m_{118} = D^2RU^2RD^2R'U^2RD^2R^2D^2$ (11) → (*ufl, fur, bdr*)

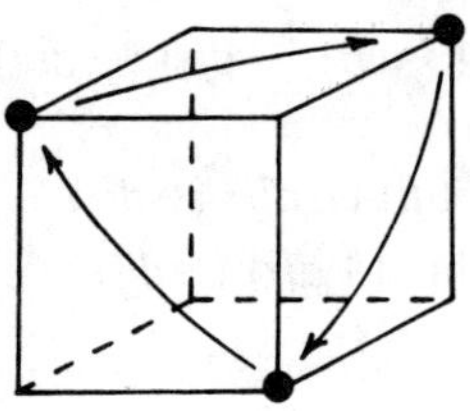

$m_{120} - m_{128}$

$m_{120} = B'R^2FR^2BR^2B'R^2F'R^2BR^2$ (12) → (*ufl, ubr, dfr*) (PK)
$m_{121} = U^2 \cdot B'DB \cdot U^2 \cdot B'D'B$ (8) → (*ufl, ubr, frd*)
$m_{122} = LD'L' \cdot U^2 \cdot LDL' \cdot U^2$ (8) → (*ufl, ubr, rdf*)
$m_{123} = F^2 \cdot L'BL \cdot F^2 \cdot L'B'L$ (8) → (*ufl, bru, dfr*)
$m_{124} = BL'B' \cdot R^2 \cdot BLB' \cdot R^2$ (8) → (*ufl, bru, frd*)
$m_{125} = RF'UB^2U'FUB^2U'R'$ (10) → (*ufl, bru, rdf*)
$m_{126} = R'UB^2U'F'UB^2U'FR$ (10) → (*ufl, rub, frd*)
$m_{127} = DB'D' \cdot F^2 \cdot DBD' \cdot F^2$ (8) → (*ufl, rub, dfr*)
$m_{128} = R^2 \cdot D'LD \cdot R^2 \cdot D'L'D$ (8) → (*ufl, rub, rdf*)

(c) Orientation preserving corner double swaps

$m_{200} = (M_R^2UM_R^2U^2)^2$ (8) → (*ufl, ubr*) (*urf, ulb*)
$m_{210} = LD^2FD' \cdot R' \cdot DF'D^2L'DF \cdot R \cdot F'D'$ (14) → (*ufl, urf*) (*ulb, ubr*) (PK)
$m_{211} = (LF^2R^2D^2RD^2 \cdot C_F^2)^2$ (12) → (*luf, rfu*) (*lbu, rub*) (PK)
$m_{211a} = F(URU'R')^3F'$ (14)
$m_{212} = B'U'BR'B^2D'BRUR'B'DB^2R$ (14) → (*ufl, urf*) (*ulb, rub*) (PK)
$m_{220} = (R^2FR^2M_D^2)^2$ (8) → (*ufl, ubr*) (*dlf, drb*)
$m_{225} = (FRF'R')^3$ (12) → (*luf, rfu*) (*frd, bdr*)
$m_{230} = (FR'F'R)^3$ (12) → (*ufl, ubr*) (*urf, dfr*)

(d) Other corner maneuvers

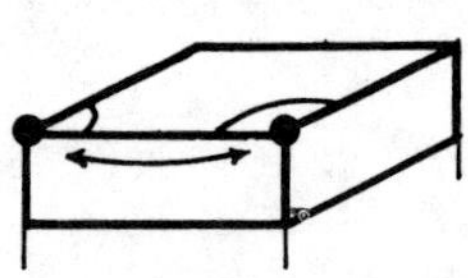

m_{300}

$m_{300} = R'DF'D^2FD'R$ (7) → (*ufl, urf*) · (*dfr, dbl*) (*df, db*) (*fl, rf*)
(Isoswap of order 2 for the upper layer, cf. Section 3.4)
$m_{310} = U^2R(UR')^2(U'R^2)^2U^2R$ (12) → (+*ufl, rub*) (−*urf, bul*)
$m_{320} = BLM_F'UM_FL'M_F'U'F' \cdot C_F$ (9) → (−*ufl, urf*) (+*ulb, ubr*) (PK)
$m_{333} = FRUR'U'F'$ (6) → (+*urf, ufl*) (−*ubr, ulb*) (*uf, ru, bu*)

m_{334} = $BRBR'U'B'$ (6) → $(-ufl, urf)$ $(+ulb, ubr)$ $(+uf, ur, lu)$ $(+ub, lb, br)$

m_{336} = $(R^2U'R^2U)^3$ (12) → $(ufl, ubr, drb, urf, dfr)$

m_{340} = $R^2U^2F^2(U^2R^2)^2F^2R^2U^2$ (10) → (ufl, ubr, dfr) (ulb, drb, dlf)

m_{350} = $M_R^2M_F^2U^2M_R^2M_F^2D^2$ (6) → (ufl, ubr) (urf, ulb) (dlf, drb) (dfr, dbl)

m_{355} = $M_F^2M_R^2UM_F^2L^2(M_D^2FM_D^2F^2)^2R^2UC_R^2$ (15) → (ufl, urf) (ulb, ubr) (dlf, dfr) (dbl, drb)

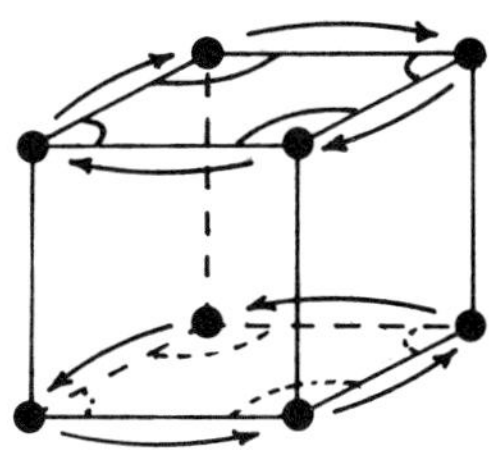

m_{360}

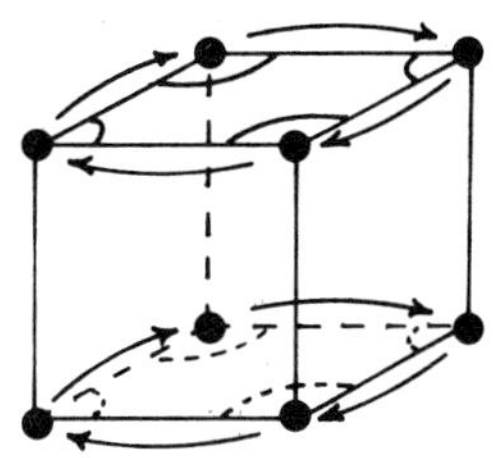

m_{365}

m_{360} = $M_R^2M_F^2UM_R^2M_F^2D$ (6) → (ufl, ulb, ubr, urf) (dlf, dfr, drb, dbl)

m_{365} = $U^2M_R^2UM_R^2U^2 \cdot M_F^2D'M_F^2$ (8) → (ufl, ulb, ubr, urf) (dlf, dbl, drb, dfr)

(e) Edge flips

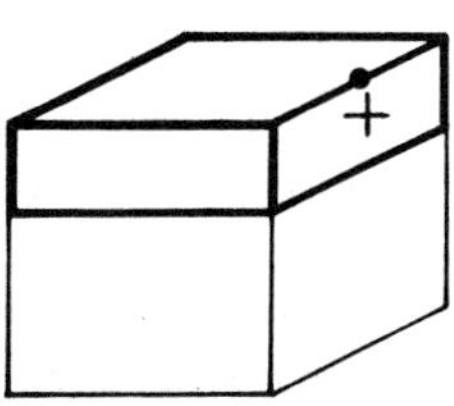

m_{400}

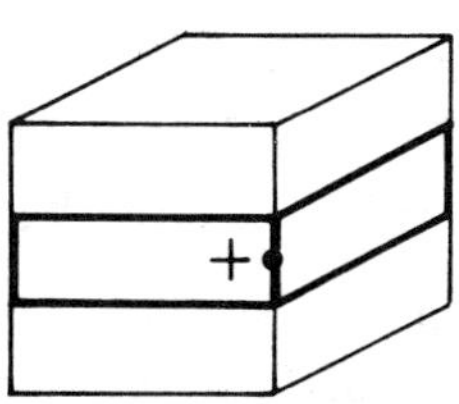

m_{401}

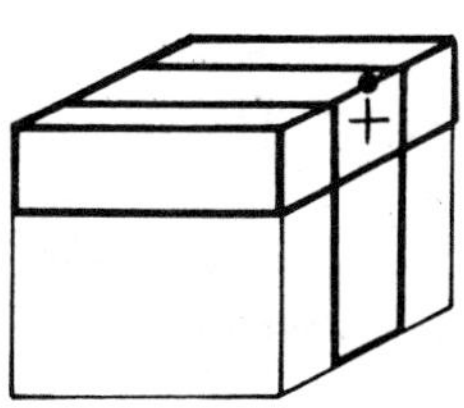

m_{402}

m_{400} = $RM_DR^2M_D^2R$ (5) → $(+ur) \cdot (+fl) \cdot (bl, rd)$ (f, l, b, r)
(isoflip for the upper layer, cf. Section 3.4)

m_{401} = $R'FD'RF'$ (5) → $(+fr) \cdot (+uf, ru, dl, db)$ $(+urf, ldb)$ $(-ufl, rub, rdf)$
(isoflip for the U-D middle layer, cf. Section 3.4)

m_{402} = $(M_DR)^4$ (8) → $(+ur) \cdot (+fl)$ $(+lb)$ $(+br)$
(isoflip of order 2 for the upper layer and for the F-B middle layer)

m_{405} = $F^2 \cdot M_DRM_DR' \cdot F^2 \cdot RM_D'RM_D'R^2$ (11) → $(+fr)$ $(+br)$ (PK)

m_{405a} = $R'FD'RF' \cdot M_D' \cdot FR'DF'R \cdot M_D$ (12)

m_{405b} = $(M_DR)^2M_DR^2 \cdot (M_D'R)^2M_D'R^2$ (12)

m_{405c} = $C_R(M_DR)^4(M_DR')^4C_R'$ (16)

$m_{406} = FR'D \cdot R^2 \cdot D'RF' \cdot R^2$ (8) $\rightarrow$ $(+fr)\ (+br) \cdot (ufl,\ dfr,\ urf,\ drb,\ ubr)$

$m_{410} = B^2 \cdot R^2M'_DRM'_DRB^2 \cdot R'M_DRM_D$ (11) $\rightarrow$ $(+fr)\ (+bl)$ (PK)

$m_{410a} = R'FD'RF' \cdot M_D^2 \cdot FR'DF'R \cdot M_D^2$ (12)

$m_{415} = RM_DR^2M_D^2R \cdot U'\ R'M_D^2R^2M'_DR' \cdot U$ (12) $\rightarrow$ $(+uf)\ (+ur)$

$m_{415a} = (M_DR)^4(M'_DF')^4$ (16)

$m_{415b} = LFR'F'L'U^2RURU'R^2U^2R$ (13)

$m_{416} = R'FRF' \cdot UF'U'F$ (8) $\rightarrow$ $(+uf)\ (+ur) \cdot (+ufl,\ dfr)\ (-urf,\ rub)$

$m_{420} = FR^2F'M_DFM_DR^2F^2M'_DFM'_D$ (11) $\rightarrow$ $(+uf)\ (+rb)$ (PK)

$m_{430} = FD^2M'_RU^2FM_D^2 \cdot F'D^2M'_RU^2F'M_D^2$ (12) $\rightarrow$ $(+fl)\ (+lb)\ (+br)\ (+rf)$

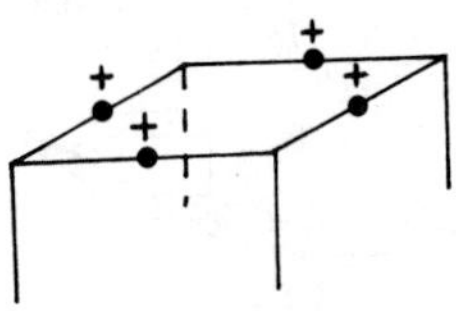

m_{435}

$m_{435} = F^2M'_RF^2M_R^2 \cdot U' \cdot M_R^2F^2M_RF^2 \cdot U$ (10) $\rightarrow$ $(+uf)\ (+ul)\ (+ub)\ (+ur)$ (H. Kraß).

$m_{435a} = UM_RU^2M'_RU^2M_R \cdot U'M'_RU^2M_RU^2M'_R$ (12) (H. Kraß)

$m_{440} = R^2M_F^2U'M_F^2R^2M'_FR^2UB'R^2M'_DR^2M_D^2BM_D^2R^2M_DM_F$ (18) $\rightarrow$ $(+uf)\ (+fr)\ (+ru)\ (+db)\ (+bl)\ (+ld)$

$m_{441} = M'_RF^2M_R^2U'M_R^2F^2M_RF^2U \cdot R'F^2M'_DF^2M_D^2RM_D^2F^2M_D$ (18) $\rightarrow$ $(+ub)\ (+br)\ (+rd)\ (+df)\ (+fl)\ (+lu)$

$m_{445} = (UR'FR)^5$ (20) $\rightarrow$ $(+uf)\ (+ul)\ (+ub)\ (+ur)\ (+fl)\ (+fd)$

$m_{470} = (M_RR^2 \cdot M_BB^2)^3$ (12) $\rightarrow$ $(+uf)\ (+ul)\ (+ub)\ (+ur)\ (+df)\ (+dl)\ (+db)\ (+dr)$

(Holding the cube with the left hand at the cube edge *lf*, we can perform M_RR^2 resp. M_BB^2 with a single quick movement of the right hand.)

$m_{480} = M_R^2(DBFRLU)^2M_R^2$ (14) $\rightarrow$ $(+uf)\ (+ul)\ (+ub)\ (+ur)\ (+fl)\ (+lb)\ (+br)\ (+rf)$

$m_{490} = ((M_RU)^4C_{ULF})^3$ (24) $\rightarrow$ $(+uf)\ (+ul)\ (+ub)\ (+ur)\ (+df)\ (+dl)\ (+db)\ (+dr)\ (+fl)\ (+lb)\ (+br)\ (+rf)$

$m_{490a} = RLFBUDRLFBU \cdot F^2M_RF^2U'M_R^2B^2M'_RB^2UM_R^2D$ (22) (H. Kraß)

(f) Edge-3-cycles

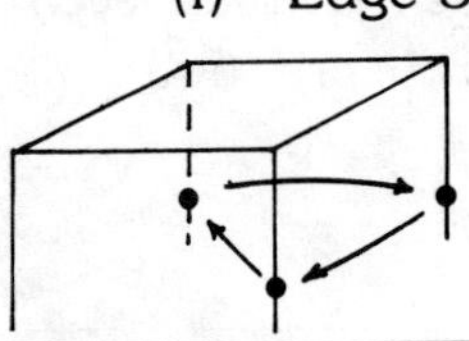

$m_{500} - m_{503}$

$m_{500} = R^2M'_DR^2M_D$ (4) → (fr, bl, br)
$m_{501} = R'M_FR'B^2RM'_FR'B^2R^2$ (9) → (fr, bl, rb)
$m_{502} = B'M_RB \cdot R^2 \cdot B'M'_RB \cdot R^2$ (8) → (fr, lb, br)
$m_{503} = B^2 \cdot RM_FR' \cdot B^2 \cdot RM'_FR'$ (8) → (fr, lb, rb)
$m_{510} = F^2UM_RU^2M'_RUF^2$ (7) → (uf, ul, ur)

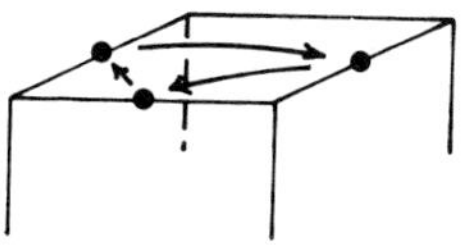

$m_{510} - m_{513}$

$m_{511} = RUR' \cdot M'_F \cdot RU'R' \cdot M_F$ (8) → (uf, ul, ru)
$m_{512} = M_F \cdot L'U'L \cdot M'_F \cdot L'UL$ (8) → (uf, lu, ur)
$m_{513} = M'_RUM_RU^2M'_RUM_R$ (7) → (uf, lu, ru)

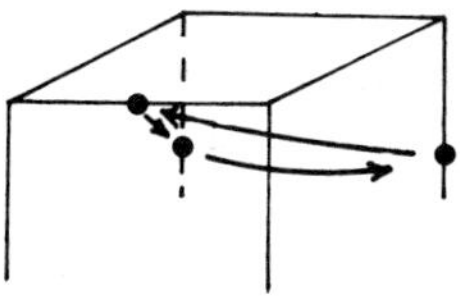

$m_{520} - m_{523}$

$m_{520} = BM_RB^2M'_RB$ (5) → (uf, lb, rb)
$m_{521} = M'_D \cdot R'U'R \cdot M_D \cdot R'UR$ (8) → (uf, lb, br)
$m_{522} = M'_D \cdot BU^2B' \cdot M_D \cdot BU^2B'$ (8) → (uf, bl, rb)
$m_{523} = M'_RB'M'_RB^2M_RB'M_R$ (7) → (uf, bl, br)

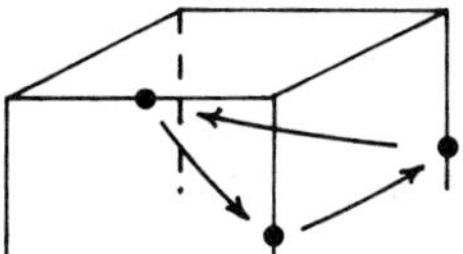

$m_{530} - m_{533}$

$m_{530} = FU^2L^2D^2BD^2L^2U^2$ (8) → (uf, rf, rb)
$m_{531} = M_D \cdot R'U'R \cdot M'_D \cdot R'UR$ (8) → (uf, rf, br)
$m_{532} = RU'R' \cdot M'_D \cdot RUR' \cdot M_D$ (8) → (uf, fr, rb)
$m_{533} = UR'M_FR^2M'_FR'U'$ (7) → (uf, fr, br)

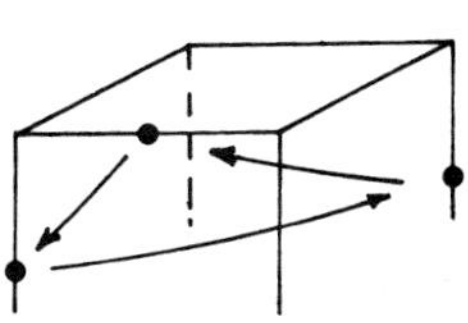

$m_{540} - m_{543}$

$m_{540} = L^2 \cdot BM_RB^2M'_RB \cdot L^2$ (7) $\rightarrow$ (uf, lf, rb)

$m_{541} = (M'_DFM_DF)^2$ (8) $\rightarrow$ (uf, lf, br)

$m_{542} = LBUM_FU^2M'_FUB'L'$ (9) $\rightarrow$ (uf, fl, rb)

$m_{543} = M_D^2 \cdot R'U'R \cdot M_D^2 \cdot R'UR$ (8) $\rightarrow$ (uf, fl, br)

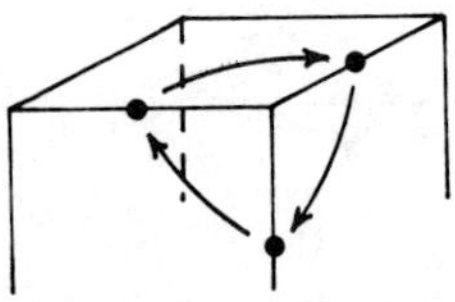

$m_{550} - m_{553}$

$m_{550} = F'RM_D^2R'U'RM_D^2R'UF$ (10) $\rightarrow$ (uf, ru, fr) (PK)

$m_{551} = U^2B'R'M_FR^2M'_FR'BU^2$ (9) $\rightarrow$ (uf, ru, rf)

$m_{551a} = ULDR \cdot F \cdot R'D'L'U' \cdot F'$ (10) (E. Rubik)

$m_{551b} = (FR)^3(F'R')^2F^2$ (11)

$m_{552} = F^2L'U'M_RU^2M'_RU'LF^2$ (9) $\rightarrow$ (uf, ur, fr)

$m_{553} = R^2D'F'M'_DF^2M_DF'DR^2$ (9) $\rightarrow$ (uf, ur, rf)

$m_{556} = R'FRF'$ (4) $\rightarrow$ (uf, ru, rf) $\cdot$ (+ufl, rub) (−urf, frd)

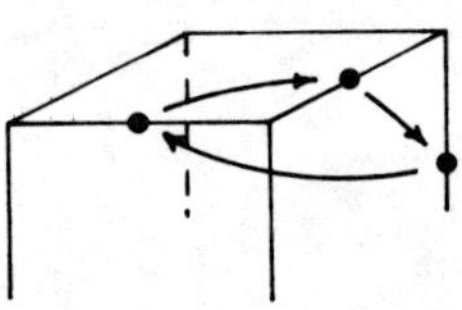

$m_{560} - m_{563}$

$m_{560} = B'M_R^2B \cdot U \cdot B'M_R^2B \cdot U'$ (8) $\rightarrow$ (uf, ur, rb)

$m_{561} = BM'_RB' \cdot U \cdot BM_RB' \cdot U'$ (8) $\rightarrow$ (uf, ur, br)

$m_{561a} = RURUR \cdot U'R'U'R'U'$ (10)

$m_{562} = R \cdot F'M_D^2F \cdot R' \cdot F'M_D^2F$ (8) $\rightarrow$ (uf, ru, rb)

$m_{563} = RBUM'_FU^2M_FUB'R'$ (9) $\rightarrow$ (uf, ru, br)

$m_{563a} = R'URBLFU'F'L'B'$ (10)

$m_{565} = U'R'UR$ (4) $\rightarrow$ (uf, ur, br) $\cdot$ (−ufl, urf) (+ubr, bdr)

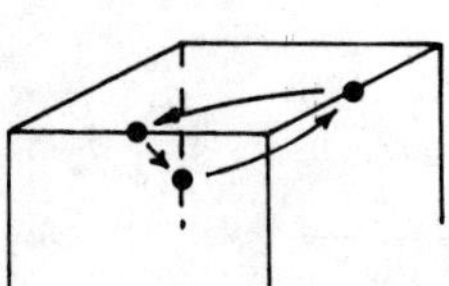

$m_{570} - m_{573}$

$m_{570} = LM_FL' \cdot U' \cdot LM'_FL' \cdot U$ (8) $\rightarrow$ (uf, lb, ur)

$m_{571} = RBM_RB^2M'_RBR'$ (7) $\rightarrow$ (uf, lb, ru)

$m_{572} = U \cdot B'M'_R B \cdot U' \cdot B'M_R B$ (8) → (uf, bl, ur)
$m_{573} = M'_R UFM_D F^2 M'_D FU'M_R$ (9) → (uf, bl, ru)
$m_{573a} = RM'_D BU^2 B'M_D BU^2 B'R'$ (10)

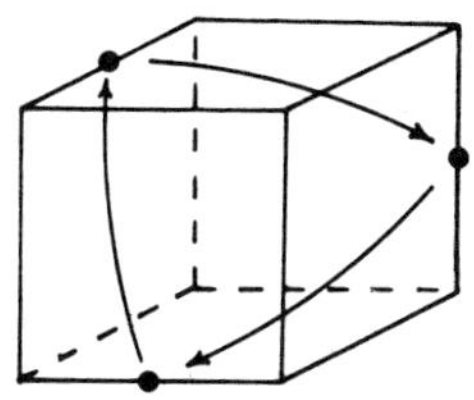

$m_{580} - m_{583}$

$m_{580} = L'M'_D RDR'M_D RD'M_R \cdot C'_R$ (9) → (ul, rb, fd)
$m_{580a} = F'M'_D RU^2 R'M_D RU^2 R'F$ (10)
$m_{581} = U'FM'_D F^2 M_D FU$ (7) → (ul, rb, df)
$m_{582} = F'RM_F R^2 M'_F RF$ (7) → (ul, br, fd)
$m_{583} = R'UM_R U^2 M'_R UR$ (7) → (ul, br, df)

(g) Orientation preserving edge double swaps

$m_{600} = (R^2 M_D^2)^2$ (4) → (fl, bl) (fr, br)
$m_{601} = M_D F^2 R^2 M'_D FM'_R F'R^2 FM_R F$ (11) → (fl, bl) (fr, rb)
$m_{602} = FR^2 M_D L^2 F \cdot M_D^2 \cdot F'R^2 M_D L^2 F'$ (11) → (fr, rb) (fl, lb)
$m_{605} = (M_R^2 M_D)^2$ (4) → (fl, br) (fr, bl)
$m_{606} = F^2 R^2 M_D RM_F RF^2 R'M'_F RM'_D$ (11) → (fl, br) (fr, lb)
$m_{607} = FR^2 M_D L^2 F' \cdot M_D^2 \cdot FR^2 M_D L^2 F'$ (11) → (fl, rb) (fr, lb) (PK)

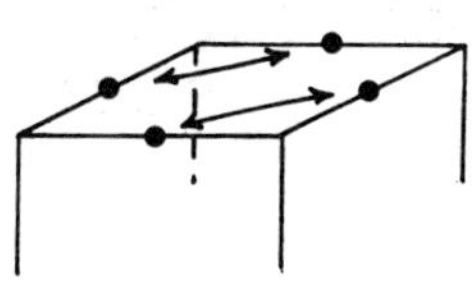

$m_{610} - m_{614}$

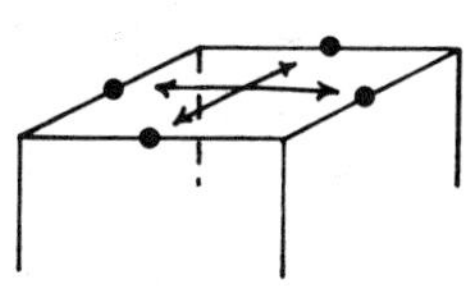

$m_{615} - m_{619}$

$m_{610} = M_F^2 D'M_R^2 DM'_R M_F^2 M_R$ (7) → (uf, ur) (ub, ul)
$m_{610a} = M_R^2 U'F^2 M_R^2 F^2 M_R^2 UM_R^2$ (8)
$m_{611} = FRUR'U'F^2 L'U'LUF$ (11) → (uf, ru) (ub, ul) (D. E. Taylor)
$m_{612} = F^2 D^2 M_R D^2 UM_R^2 UM_R U^2 F^2$ (10) → (uf, ru) (ub, lu) (PK)
$m_{612a} = RLF(R^2 U^2)^3 F'L'R'$ (12) (D. Singmaster)
$m_{615} = M_R^2 UM_R^2 U^2 M_R^2 UM_R^2$ (7) → (uf, ub) (ul, ur)
$m_{615a} = RLU^2 R'L' \cdot F'B'U^2 FB$ (10) (PK)

$m_{616} = FM_RF \cdot UM_R^2U^2M_R^2U \cdot F'M_R'F'$ (11) → (*uf, bu*) (*ul, ur*)

$m_{616a} = F'UR' \cdot UF^2U^2F^2U^2F^2U \cdot RU'F$ (13) (PK)

$m_{617} = M_R^2F^2M_R'F^2UF^2M_RF^2M_R^2U'$ (10) → (*uf, bu*) (*ul, ru*) (H. Kraß)

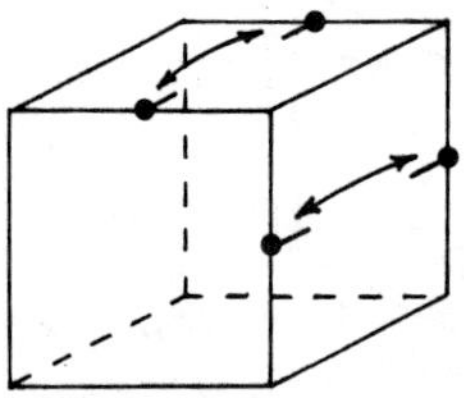

m_{620}

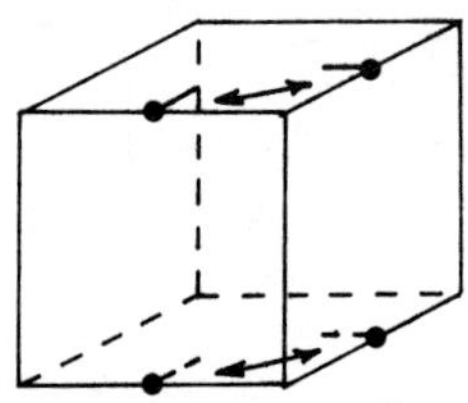

m_{625}

$m_{620} = (R^2U^2)^3$ (6) → (*rf, rb*) (*uf, ub*)
(Holding the cube with the left thumb and forefinger by the edge cubies *fu* and *bu* and with the right thumb and forefinger by the edge cubies *fr* and *br*, we can perform m_{620} without letting go these cubies: "D. Singmaster's magic grip".)

$m_{625} = D^2M_FM_R'D'M_RD^2M_F'D$ (8) → (*uf, ur*) (*df, dr*)

$m_{630} = M_R'D^2M_RD'M_FD^2M_F'D$ (8) → (*uf, df*) (*ur, db*)

(h) Other edge maneuvers

$m_{700} = R^2M_DFM_R'F'R^2FM_RFM_D'F^2$ (11) → (+*fr, br*) (+*fl, bl*)

$m_{701} = L^2F^2M_D'FM_R'FL^2F'M_RFM_D$ (11) → (+*fr, rb*) (+*fl, bl*)

$m_{703} = F^2R^2M_DFM_R'F'R^2FM_RFM_D'$ (11) → (+*fr, bl*) (+*fl, br*)

$m_{706} = F^2M_RU'M_R'U'M_RUM_R'UF^2$ (10) → (+*ul, ub*) (+*ur, uf*)

$m_{707} = B'F^2 \cdot R'U'M_RURUM_R'U' \cdot F^2B$ (12) → (+*uf, ur*) (+*ul, ub*) (PK)

$m_{707a} = RBLU'L'UB'R' \cdot B'U'R'URB$ (14) (M. Henze)

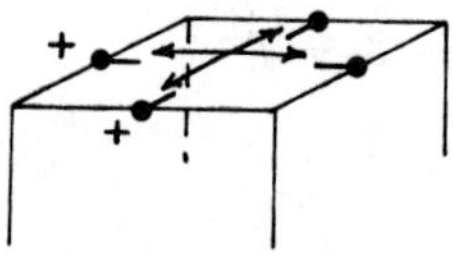

m_{710})

$m_{710} = FU^2F'U'L'B'U^2BUL$ (10) → (+*uf, ub*) (+*ul, ur*)

$m_{713} = UBU^2B'R'U'B'R^2BR$ (10) → (+*rf, rb*) (+*ub, uf*) (M. B. Thistlethwaite)

$m_{716} = FDR^2D'R'F'D'F^2DR$ (10) → (+*uf, rb*) (+*ur, lf*)

$m_{720} = UM_RU'M_R' \cdot U'M_RUM_R'$ (8) → (+*uf, ul*) (+*ub, df*)

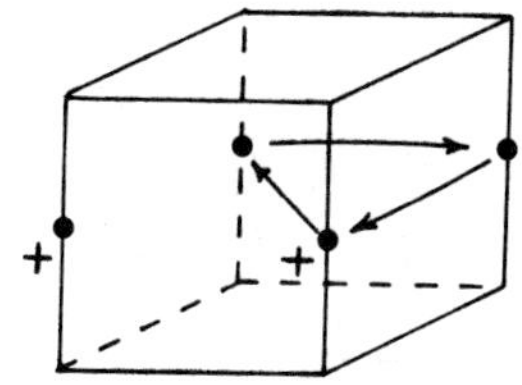

m_{730}

$m_{730} = FM_DF'M_D^2L^2F'M_DFL^2$(9) → (+*fl*) (+*fr, bl, br*) (M. and T. Schweizer)

$m_{731} = F^2LM_DL'F^2M_D^2L'M_DL$(9) → (+*fl*) (+*fr, bl, rb*) (M. and T. Schweizer)

$m_{733} = L^2F'M_R'FL^2F'M_RF'M_D'F^2M_D$ (11) → (+*fl*) (+*fr, lb, rb*)

$m_{750} = D'M_R'DM_R$ (4) → (*uf, rd, dl, db, fd*)

$m_{755} = UM_R^2U'M_R^2$ (4) → (*uf, db, ur*) (*ub, df, ul*)

$m_{756} = R^2U^2F^2U'M_F^2UF^2U^2M_F^2R^2$ (10) → (*uf, ul, ur*) (*df, dr, dl*)

$m_{757} = R^2B^2UM_F^2U'F^2L^2C_F^2$ (7) → (*uf, ul, ur*) (*db, dr, dl*)

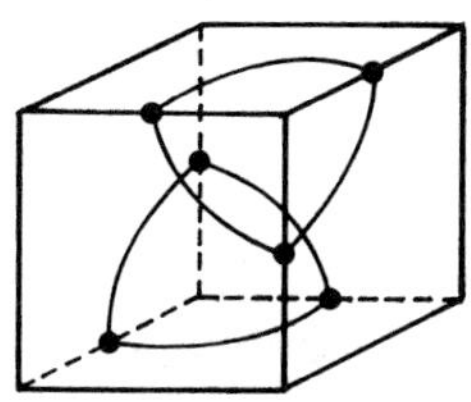

$m_{758} - m_{761}$

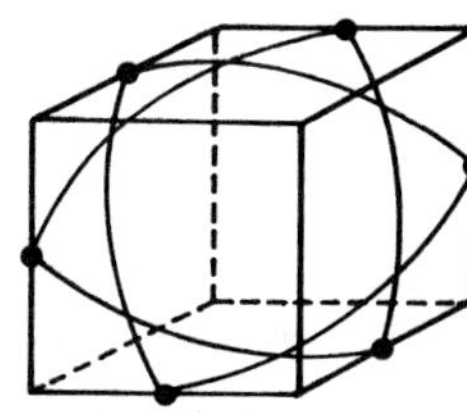

$m_{762} - m_{765}$

$m_{758} = M_RD'L^2B^2F' \cdot UM_R^2U'M_R^2 \cdot FB^2L^2DM_R'$ (14) → (*uf, ru, fr*) (*db, ld, bl*) (PK)

$m_{759} = RLF^2U' \cdot BM_R^2B'M_R^2 \cdot UF^2R'L'$ (12) → (*uf, ru, fr*) (*db, bl, ld*) (PK)

$m_{760} = D'L'F \cdot DBM_DF'R'FM_D'B' \cdot F'LD$ (14) → (+*uf, ur, fr*) (+*db, dl, bl*) (R. aus dem Spring)

$m_{762} = D'F^2R' \cdot M_FD^2UM_F'D^2U' \cdot RF^2D$ (12) → (*ul, rb, fd*) (*ub, rd, fl*) (PK)

$m_{763} = M_R'FM_R^2U'M_R'U^2M_F'U'M_FF^2M_D'FM_D$ (13) → (*ul, rb, fd*) (*ub, fl, rd*) (R. aus dem Spring)

$m_{764} = RUF^2U'BU^2F'B'R'L'U^2LU'R^2UF$ (16) → (+*ul, br, fd*) (+*ub, dr, fl*) (PK)

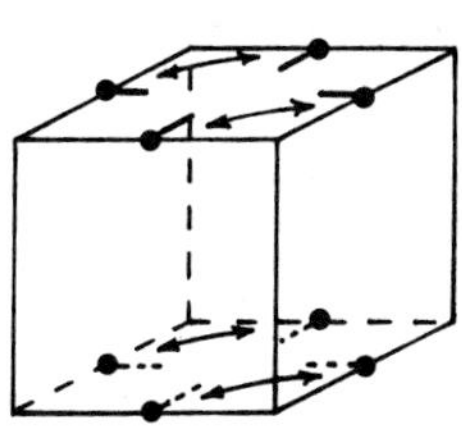

m_{770}

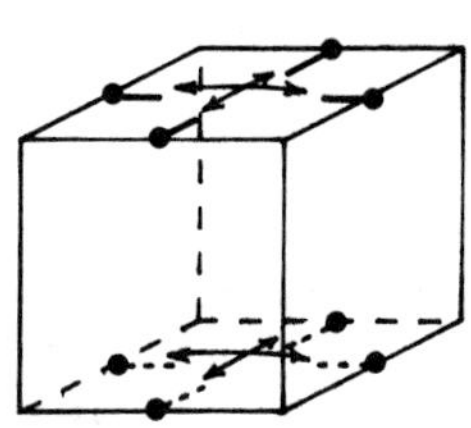

m_{771}

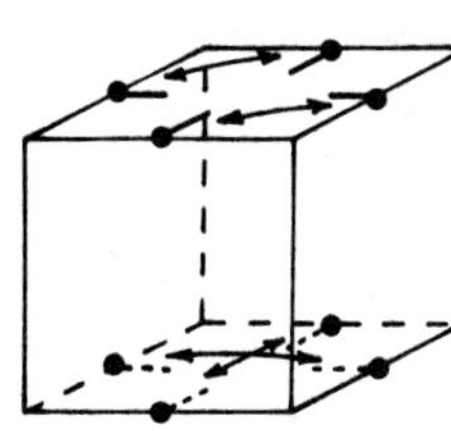

m_{772}

$m_{765} = L^2 \cdot F^2UR^2M_F^2R^2M_F^2U'F^2 \cdot B'M_R'BR'BM_RB'R \cdot L^2$ (18) → (+*ul, br, fd*) (+*ub, lf, rd*) (PK)

$m_{766} = M_F^2 D' M_R^2 D' M_R' M_F^2 M_R D^2$ (8) → (uf, ul, ub, ur) (dl, dr)
$m_{768} = M_F^2 M_R^2 D^2 M_F^2 D' M_F' D^2 M_F D' M_R^2 D^2$ (11) → (uf, ul, ur) (df, db) (dl, dr)
$m_{770} = M_R^2 M_F^2 U M_R^2 U^2 M_F^2 U$ (7) → (uf, ur) (ub, ul) (df, dr) (db, dl)
$m_{771} = (M_R^2 M_F^2 U^2)^2$ (6) → (uf, ub) (ul, ur) (df, db) (dl, dr)

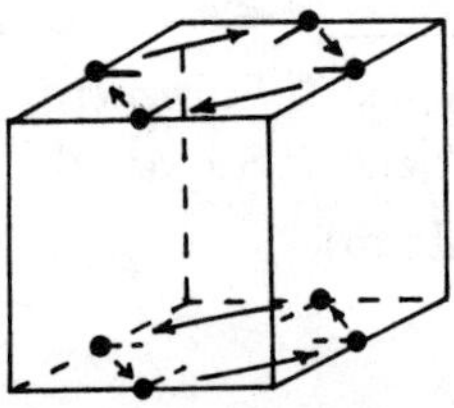

m_{780}

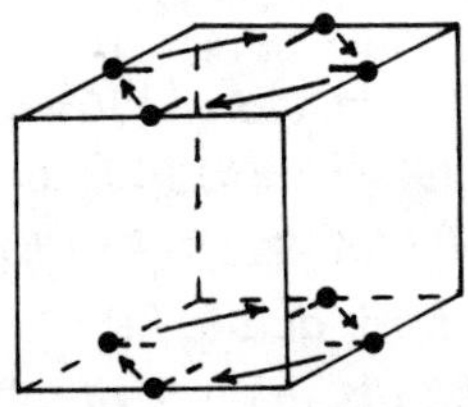

m_{781}

$m_{772} = M_F^2 D M_R^2 D M_R M_F^2 M_R D^2 M_R^2$ (9) → (uf, ur) (ub, ul) (df, db) (dl, dr)
$m_{773} = D^2 M_R M_F^2 M_R' U^2 C_U^2$ (5) → (fr, bl) (fl, br) (ul, dl) (ur, dr)
$m_{774} = RLM_D^2 R'L' M_D^2$ (6) → (fr, bl) (fl, br) (ul, ur) (dl, dr)
$m_{775} = M_R' M_D^2 M_R M_F^2$ (4) → (fr, bl) (fl, br) (ul, dr) (ur, dl)
$m_{780} = D M_R^2 M_F^2 D' M_R^2 M_F^2$ (6) → (uf, ul, ub, ur) (df, dr, db, dl)
$m_{781} = M_R^2 U^2 M_F^2 D M_F^2 U^2 M_R^2 D'$ (8) → (uf, ul, ub, ur) (df, dl, db, dr)
$m_{782} = M_R M_D M_F M_D'$ (4) → (fu, ub, bd, df) (lu, ur, rd, dl)
$m_{790} = M_R^2 M_F^2 M_D^2$ (3) → (uf, db) (ub, df) (ul, dr) (ur, dl) (fl, br) (fr, bl)
$m_{791} = F^2 R^2 U^2 M_F^2 D^2 R^2 F^2 \cdot C_F$ (7) → (uf, ub) (df, db) (ru, rd) (lu, ld) (fl, fr) (bl, br)

(i) Face cubie turns ("invisible operations")

$m_{800} = (URLU^2R'L')^2$ (12) → (++u) (M. B. Thistlethwaite)
$m_{801} = RLU^2R^2M_F'M_R' \cdot F \cdot M_R M_F R^2 U^2 L'R' \cdot U$ (14) → (+u) (+r) (PK)
$m_{801a} = (U'R^2U^2R'U^2R^2)^3$ (18)
$m_{802} = M_R' M_D' M_R \cdot U' \cdot M_R' M_D M_R \cdot U$ (8) → (+u) (−r)
$m_{803} = M_R' M_D' M_R \cdot U^2 \cdot M_R' M_D M_R \cdot U^2$ (8) → (++u) (++r)
$m_{804} = M_F^2 M_R' B M_R M_F^2 R^2 D^2 R'L'DRLD^2R^2$ (14) → (+u) (+d) (PK)
$m_{805} = M_R' M_D^2 M_R \cdot U' \cdot M_R' M_D^2 M_R \cdot U$ (8) → (+u) (−d)
$m_{806} = (M_R' M_D^2 M_R U^2)^2$ (8) → (++u) (++d)

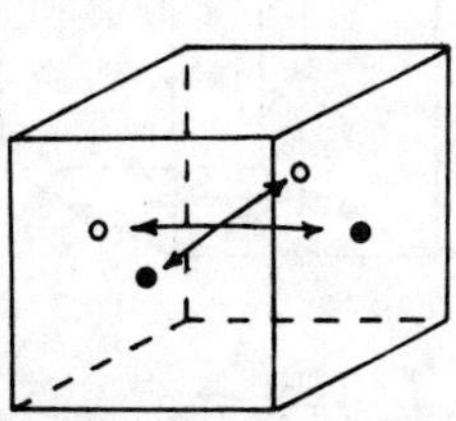

m_{898}

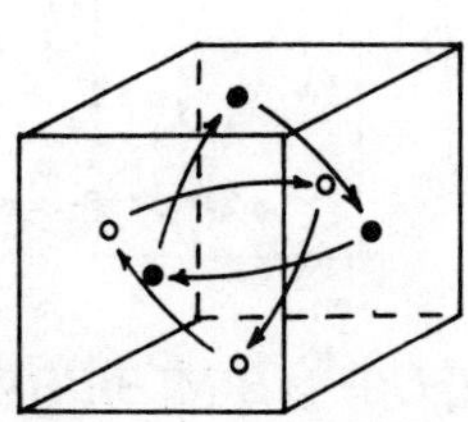

m_{899}

(j) Other face cubie maneuvers

$m_{898} = M_R^2 M_D' M_R^2 M_D$ (4) → (*f, b*) (*l, r*)

$m_{899} = M_R M_F' M_R' M_F$ (4) → (*u, r, f*) (*d, l, b*)

(k) Corner and edge cubies

$m_{900} = RF^2R'BRB^2D^2FLF'D^2BF^2R'$ (14) → (*ufl, urf*) (*ul, ur*) (PK)

$m_{901} = F^2U'F^2UF^2R^2UR^2D'F^2D$ (11) → (*ufl, urf*) (*uf, ub*)

$m_{902} = L^2B'U'BL^2F'DF'D'F^2$ (10) → (*ufl, urf*) (*uf, ur*) (PK)

$m_{903} = B^2RBUB'U'B'UR'B'RU'R'B'$ (14) → (*ufl, urf*) (*ul, ub*) (PK)

$m_{904} = RD^2 \cdot UL'ULU^2F^2DRD'F^2 \cdot D^2R'$ (14) → (*ufl, ubr*) (*uf, ur*)

$m_{905} = B'R'URB'D'RFRF'R^2DB^2$ (13) → (*ufl, ubr*) (*uf, ul*)

$m_{906} = LD' \cdot LBL^2D^2RFR'D^2LB' \cdot DL'$ (14) → (*ufl, ubr*)(*uf, ub*) (PK)

$m_{915} = RUR'U'F'U'F$ (7) → (+*urf, ubr*) (−*ulb*) (*ur, ul, rf, bu*)

$m_{920} = F'U'BU^2B'URU^2R'F$ (10) → (*ulb, rbd, fdl*) (*ul, rb, fd*) (R. Walker)

$m_{921} = U'R^2D'B'DR'D'BDR'U$ (11) → (*ufl, rub, frd*) (*uf, ru, fr*) (PK)

$m_{925} = LD'FD^2F'DRD^2R'L'$ (10) → (*ufl, urf, ubr*) (*fl, rf, br*) (PK)

(1) Corner and face cubies

(m) Edge and face cubies

$m_{970} = RFM_RDR \cdot M_F' \cdot R'D'M_R'F'R'$ (11) → (*uf, ul, ub, ur*) (*f, l, b, r*) (R. aus dem Spring)

$m_{980} = (L^2M_D)^2$ (4) → (*fr, bl, br*) (*f, b*) (*l, r*)

$m_{981} = M_R' M_D^2 M_R$ (3) → (*fr, bl*) (*fl, br*) (*u, d*) (*l, r*)

(n) Maneuvers of theoretical interest

$m_{990} = RL'F^2B^2RL' \cdot U \cdot LR'B^2F^2LR'$ (13) $\sim D$ (D. J. Benson)

$m_{991} = UBLUL'U'B'$ (7) → (−*ufl, ubr*) (+*ulb*) (+*ub, ur*) (+*ul*)

$m_{992} = R^2FLD'R'$ (5) → (*urf, bdr, lfd, ldb, lbu, luf, frd*) (*uf, rd, rf, ru, fd, db, br, ld, lb, lu, lf*)

m_{991} and m_{992} cause operations which generate Rubik's group G (F. Barnes)

$m_{996} = RU^2D'BD'$ (5) → (−*ufl, lbu, rfu*) (+*ubr, fdl, dfr, rbd, ldb*) (+*uf, lb, dr, fr, ul, ur, bu*) (+*dl, rb*) (*df, db*)

m_{996} causes an operation of the maximal order 1260 (J. B. Butler)

$m_{997} = RC_U$ (1) → (+*ufl, ulb, ubr, rdf*) (+*urf*) (+*dlf, dbl, drb*) (+*uf, ul, ub, ur, rf*) (+*df, dl, db, dr, lf, bl, rb*) (*f, l, b, r*)

m_{997} causes an 1-operation of the maximal order 1260

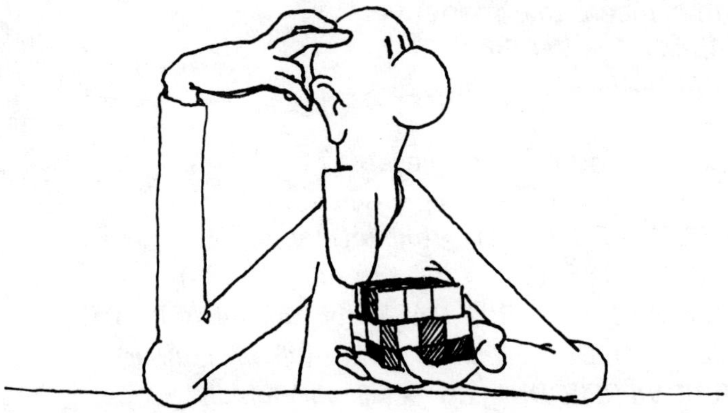

BIBLIOGRAPHY

(a) Books

Bandelow, C. & Klerings, P.: International Handbook of Rubik's Cube, 1982

Bindweed, W. C. et al.: Not another Cube book! 92 pp. Ballantine Books, New York, and Pan Books, London & Sydney 1981

Bossert, P.: You can do the Cube. 112 pp. Penguin 1981

Deledicq, A. & Touchard, J.-B.: Le Cube Hongrois—mode d'emploi. 76 pp. 1981 (Institut de Recherche pour l'Enseignement des Mathématiques, 2. Place Jussieu, 75005 Paris, France)

Endl, K.: Rubik's Rätsel des Jahrhunderts. 112 pp. Würfel-Verlag, Giessen 1981 (English translation: Rubik's Cube made simple. 112 pp. Würfel-Verlag, Giessen 1981)

Endl, K.: Die Pyramide. 32 pp. Würfel-Verlag, Giessen 1981 (English translation: The Pyramid. 32 pp. Würfel-Verlag, Giessen 1981)

Hammond, N.: How to solve the Cube in 37 seconds. 20 pp. 1981 (DRG Blackhall, Longmoor Lane, Breaston, Derby, DE7 3BQ, Great Britain)

Klemm, M. F.: Der Rubiksche Würfel. 52 pp. 1981 (POB 2733, D-6500 Mainz)

Kośniowski, C.: Master the Rubik Cube easily. 30 pp. Kornworthy Publications. Eastleigh 1981

Meier, K.: Puzzlespass mit dem Rubik's Cube. 64 pp. 1981 (Friedrich-Lau-Str. 28, D-4000 Düsseldorf 30)

Meier, K.: The small Magic Tetrahedron. 18 pp. 1981 (Friedrich-Lau-Str. 28, D-4000 Düsseldorf 30)

Mrowka, M. and Weber, W. J. (Ed.): Mathematische Überlegungen zum Würfel. Der Mathematik-Unterricht 6/1981

Mrowka, M. & Weber, W. J.: Zaubertrum, Teufelstonne und magische Pyramide. 128 pp. Falken-Verlag, Niedernhausen 1981

Nourse, J. G.: The simple solution to Rubik's Cube. 64 pp. Bantam Books, Toronto-New York-London-Sydney 1981

ROMAX: Hob Kill die Hut Uhr (Zauberwürfel-Handbuch). With English Supplement. 60 pp. 1981 (Max Heinrich Rostock, Augusta-Anlage 10, D-6800 Mannheim)

Rubik, E. et al.: A buvös kocka. 206 pp. Muszaki Konyvkiado, Budapest 1981

Singmaster, D.: Notes on Rubik's 'Magic Cube'. 5. ed. 75 pp. Enslow Publishers, Hillside, New Jersey 1980

Taylor, D.: Mastering Rubik's Cube. 96 pp. Penguin 1980

Trajber, J.: Der Würfel. 144 pp. Falken-Verlag, Niedernhausen 1981

Trajber, J.: Der Würfel für Fortgeschrittene. 144 pp. Falken-Verlag, Niedernhausen 1981

Turner-Smith, R. F.: The amazing Pyraminx. 192 pp. Mèffert Novelties Ltd, Hong Kong 1981

Warusfel, A.: Réussir le Rubik's Cube. 192 pp. Denoël, Paris 981 (Spanish translation: El Cubo Magico de Rubik. 143 pp. Altalena Editores, Madrid 1981)

Werneck, T.: Der Zauberwürfel. 160 pp. Heyne, München 1981

Werneck, T.: Der Zauberwürfel für Könner. 160 pp. Heyne München 1982

Werneck, T.: Zauberpyramide, Teufelstonne, Tower, Trikki 4. 208 pp. Heyne, München 1981 (English translation: Mastering the Magic Pyramid. Evans Brothers, London 1981)

Varasano, J.: Jeff conquers the Cube in 45 seconds—and you can too! 52 pp. Day Books, New York 1981

Zales, J. & Stevens, J.: 101 uses for a dead cube. 96 pp. Tom Doherty Associates, New York 1981

(b) Papers

Bandelow, C.: Rubik's Zauberwürfel. Jahrbuch Überblicke Mathematik 1981, 227-244. Bibliographisches Institut, Mannheim 1981

Becker, J.: Über die Anzahl der verschiedenen "Gesichter" von Rubik's Cube. Praxis der Mathematik 9/1981, 278-280

Berlekamp, E. R. & Conway, J. H. & Guy, R. K,: The Hungarian Cube—Buvös Kocka. Chapter 24 of "Winning Ways," London, Academic Press 1981

Bild der Wissenschaft, 11/1980: 174-177, 12/1980: 180-187, 4/1981: 99-106, 8/1981: 112-115, 9/1981: 132, 10/1981: 224, 11/1981: 197-200, 12/1981: 179-185

Celmins, U.: The Hungarian Magic Cube Puzzle. Annals of Discrete Mathematics 9(1980), 13-20

Dauphin, M.: Un cube pas comme les autres. Mathématique et Pédagogie 24 (Nov.-Dec. 1979), 23-29

Golomb, W.: Rubik's Cube and a model of quark confinement. Amer. Journal of Physics, Nov. 1981

Halberstadt, E.: Cube hongrois et théorie des groupes. Pour la Science (French edition of Scientific American), Aug. 1980, 23-26

Hofstadter, D. R.: The Magic Cube's cubies . . . Scientific American 3/1981 (German translation in Spektrum der Wissenschaft 5/1981, Spanish translation in Investigacion y Ciencia 5/1981)

Mrowka, M & Weber, J.: Der Ungarische Würferl im Unterricht. Praxis der Mathematik 5/1981, 129-137

Neumann, P. M. et al.: The group theory of the Hungarian Magic Cube. Chapter 19 of "Groups and Geometry". 9 pp. Math. Inst. of Oxford 1980

Ollerenshaw, D. K.: The Hungarian Magic Cube. Bull. of the Inst. of Math. and its Appl. 16(1980), 86-92

Der Spiegel, 19.1.1981, 181-184: Erfreue dich der Symmetrie (A. Kunkel)

Weber, W. J.: Die didaktischen Möglichkeiten des Ungarischen Würfels. Praxis der Mathematik 2/1981, 33-38

Weber, W. J.: Ein Problem der räumlichen Geometrie und seine Lösung mit Hilfe des Ungarischen Würfels. math. did. 1980, 219-224

(c) Patents

Rubik, E.: Hungarian Patent Nr. 170062 dated 30.1.1975, 28.10.1976, 31.12.1977 (3x3x3- and 2x2x2-Cube)

Ishige, T.: Japanese Patent Nr. 55-8192 dated 12.10.1976, 26.4.1978, 3.3.1980 (3x3x3-Cube)

Ishige, T.: Japanese Patent Nr. 55-8193 dated 12.3.1977, 4.10.1978, 3.3.1980 (2x2x2-Cube)

Ishige, T.: Japanese Patent Nr. 55-3956 dated 29.3.1977, 21.10.1978, 28.1.1980 (3x3x3-Cube)

INDEX